T0092537

Contributions to Statistics

V. Fedorov/W.G. Müller/I. N. Vuchkov (Eds.)
Model-Oriented Data Analysis,
XII/248 pages, 1992

J. Antoch (Ed.)
Computational Aspects of Model Choice,
VII/285 pages, 1993

W. G. Müller/H. P. Wynn/A. A. Zhigljavsky (Eds.)
Model-Oriented Data Analysis,
XIII/287 pages, 1993

P. Mandl/M. Hušková (Eds.)
Asymptotic Statistics,
X/474 pages, 1994

P. Dirschedl/R. Ostermann (Eds.)
Computational Statistics,
VII/553 pages, 1994

C. P. Kitsos/W. G. Müller (Eds.)
MODA 4 – Advances in Model-Oriented Data Analysis,
XIV/297 pages, 1995

H. Schmidli
Reduced Rank Regression,
X/179 pages, 1995

W. Härdle/M. G. Schimek (Eds.)
Statistical Theory and Computational Aspects of Smoothing,
VIII/265 pages, 1996

S. Klinke
Data Structures for Computational Statistics,
VIII/284 pages, 1997

A. C. Atkinson/L. Pronzato/H. P. Wynn (Eds.)
MODA 5 – Advances in Model-Oriented Data
Analysis and Experimental Design
XIV/300 pages, 1998

M. Moryson
Testing for Random Walk
Coefficients in Regression
and State Space Models
XV/317 pages, 1998

S. Biffignandi (Ed.)
Micro- and Macrodata of Firms
XII/776 pages, 1999

Wolfgang Härdle · Hua Liang
Jiti Gao

Partially Linear Models

With 17 Figures
and 11 Tables

Physica-Verlag
A Springer-Verlag Company

Series Editors
Werner A. Müller
Martina Bihn

Authors
Professor Dr. Wolfgang Härdle
Institut für Statistik und Ökonometrie
Humboldt-Universität zu Berlin
Spandauer Straße 1
10178 Berlin, Germany
Email: haerdle@wiwi.hu-berlin.de

Professor Dr. Hua Liang
Frontier Science & Technology
Research Foundation
Harvard School of Public Health
1244 Boylston Street
Cestnut Hill, MA 02467, USA
Email: hliang@fstrf.dfci.harvard.edu

Professor Dr. Jiti Gao
Department of Mathematics and Statistics
The University of Western Australia
Nedlands WA 6907, Australia
Email: jiti@maths.uwa.edu.au

This book is also available as e-book on http://www.i-xplore.de

---

ISSN 1431-1968
ISBN 3-7908-1300-1 Physica-Verlag Heidelberg New York

Cataloging-in-Publication Data applied for
Die Deutsche Bibliothek – CIP-Einheitsaufnahme
Härdle, Wolfgang: Partially linear models: with 11 tables/Wolfgang Härdle; Hua Liang; Jiti Gao. – Heidelberg: Physica-Verl., 2000
(Contributions to statistics)
ISBN 3-7908-1300-1

This work is subject to copyright. All rights are reserved, whether the whole or part of the material is concerned, specifically the rights of translation, reprinting, reuse of illustrations, recitation, broadcasting, reproduction on microfilm or in any other way, and storage in data banks. Duplication of this publication or parts thereof is permitted only under the provisions of the German Copyright Law of September 9, 1965, in its current version, and permission for use must always be obtained from Physica-Verlag. Violations are liable for prosecution under the German Copyright Law.

Physica-Verlag Heidelberg New York
a member of BertelsmannSpringer Science+Business Media GmbH

© Physica-Verlag Heidelberg 2000
Printed in Germany

The use of general descriptive names, registered names, trademarks, etc. in this publication does not imply, even in the absence of a specific statement, that such names are exempt from the relevant protective laws and regulations and therefore free for general use.

Softcover design: Erich Kirchner, Heidelberg

SPIN 10764630 88/2202-5 4 3 2 1 0 – Printed on acid-free paper

PREFACE

In the last ten years, there has been increasing interest and activity in the general area of partially linear regression smoothing in statistics. Many methods and techniques have been proposed and studied. This monograph hopes to bring an up-to-date presentation of the state of the art of partially linear regression techniques. The emphasis of this monograph is on methodologies rather than on the theory, with a particular focus on applications of partially linear regression techniques to various statistical problems. These problems include least squares regression, asymptotically efficient estimation, bootstrap resampling, censored data analysis, linear measurement error models, nonlinear measurement models, nonlinear and nonparametric time series models.

We hope that this monograph will serve as a useful reference for theoretical and applied statisticians and to graduate students and others who are interested in the area of partially linear regression. While advanced mathematical ideas have been valuable in some of the theoretical development, the methodological power of partially linear regression can be demonstrated and discussed without advanced mathematics.

This monograph can be divided into three parts: part one–Chapter 1 through Chapter 4; part two–Chapter 5; and part three–Chapter 6. In the first part, we discuss various estimators for partially linear regression models, establish theoretical results for the estimators, propose estimation procedures, and implement the proposed estimation procedures through real and simulated examples.

The second part is of more theoretical interest. In this part, we construct several adaptive and efficient estimates for the parametric component. We show that the LS estimator of the parametric component can be modified to have both Bahadur asymptotic efficiency and second order asymptotic efficiency.

In the third part, we consider partially linear time series models. First, we propose a test procedure to determine whether a partially linear model can be used to fit a given set of data. Asymptotic test criteria and power investigations are presented. Second, we propose a Cross-Validation (CV) based criterion to select the optimum linear subset from a partially linear regression and establish a CV selection criterion for the bandwidth involved in the nonparametric kernel estimation. The CV selection criterion can be applied to the case where the observations fitted by the partially linear model (1.1.1) are independent and identically distributed (i.i.d.). Due to this reason, we have not provided a separate chapter to discuss the selection problem for the i.i.d. case. Third, we provide recent developments in nonparametric and semiparametric time series regression.

This work of the authors was supported partially by the Sonderforschungsbereich 373 "Quantifikation und Simulation Ökonomischer Prozesse". The second author was also supported by the National Natural Science Foundation of China and an Alexander von Humboldt Fellowship at the Humboldt University, while the third author was also supported by the Australian Research Council. The second and third authors would like to thank their teachers: Professors Raymond Carroll, Guijing Chen, Xiru Chen, Ping Cheng and Lincheng Zhao for their valuable inspiration on the two authors' research efforts. We would like to express our sincere thanks to our colleagues and collaborators for many helpful discussions and stimulating collaborations, in particular, Vo Anh, Shengyan Hong, Enno Mammen, Howell Tong, Axel Werwatz and Rodney Wolff. For various ways in which they helped us, we would like to thank Adrian Baddeley, Rong Chen, Anthony Pettitt, Maxwell King, Michael Schimek, George Seber, Alastair Scott, Naisyin Wang, Qiwei Yao, Lijian Yang and Lixing Zhu.

The authors are grateful to everyone who has encouraged and supported us to finish this undertaking. Any remaining errors are ours.

Berlin, Germany — Wolfgang Härdle
Texas, USA and Berlin, Germany — Hua Liang
Perth and Brisbane, Australia — Jiti Gao

CONTENTS

1 INTRODUCTION

1.1 Background, History and Practical Examples

A partially linear regression model of the form is defined by

$$Y_i = X_i^T\beta + g(T_i) + \varepsilon_i, i = 1, \ldots, n \tag{1.1.1}$$

where $X_i = (x_{i1}, \ldots, x_{ip})^T$ and $T_i = (t_{i1}, \ldots, t_{id})^T$ are vectors of explanatory variables, (X_i, T_i) are either independent and identically distributed (i.i.d.) random design points or fixed design points. $\beta = (\beta_1, \ldots, \beta_p)^T$ is a vector of unknown parameters, g is an unknown function from $\mathbb{R}^d$ to $\mathbb{R}^1$, and $\varepsilon_1, \ldots, \varepsilon_n$ are independent random errors with mean zero and finite variances $\sigma_i^2 = E\varepsilon_i^2$.

Partially linear models have many applications. Engle, Granger, Rice and Weiss (1986) were among the first to consider the `partially linear model` (1.1.1). They analyzed the relationship between temperature and electricity usage.

We first mention several examples from the existing literature. Most of the examples are concerned with practical problems involving partially linear models.

Example 1.1.1 *Engle, Granger, Rice and Weiss (1986) used data based on the monthly electricity sales y_i for four cities, the monthly price of electricity x_1, income x_2, and average daily temperature t. They modeled the electricity demand y as the sum of a smooth function g of monthly temperature t, and a linear function of x_1 and x_2, as well as with* 11 *monthly dummy variables $x_3, \ldots, x_{13}$. That is, their model was*

$$\begin{aligned} y &= \sum_{j=1}^{13} \beta_j x_j + g(t) \\ &= X^T\beta + g(t) \end{aligned}$$

where g is a smooth function.

In Figure 1.1, the nonparametric estimates of the weather-sensitive load for St. Louis is given by the solid curve and two sets of parametric estimates are given by the dashed curves.

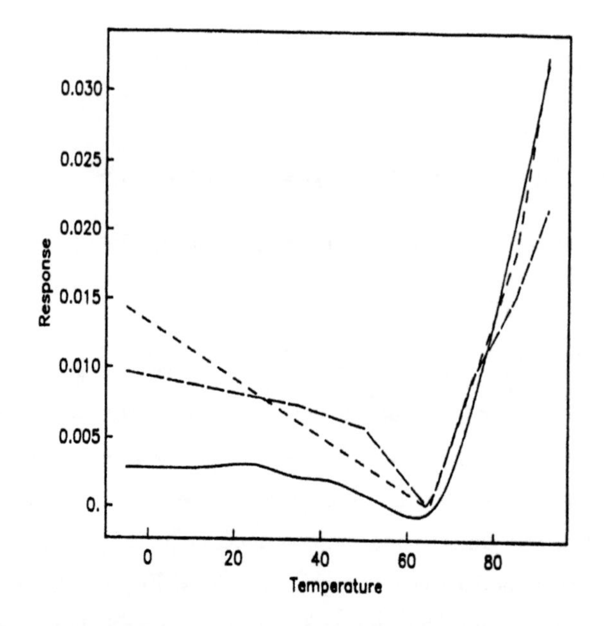

FIGURE 1.1. Temperature response function for St. Louis. The nonparametric estimate is given by the solid curve, and the parametric estimates by the dashed curves. From Engle, Granger, Rice and Weiss (1986), with permission from the Journal of the American Statistical Association.

Example 1.1.2 *Speckman (1988) gave an application of the partially linear model to a mouthwash experiment. A control group ($X = 0$) used only a water rinse for mouthwash, and an experimental group ($X = 1$) used a common brand of analgesic. Figure 1.2 shows the raw data and the partially kernel regression estimates for this data set.*

Example 1.1.3 *Schmalensee and Stoker (1999) used the partially linear model to analyze household gasoline consumption in the United States. They summarized the modelling framework as*

$$\begin{aligned} LTGALS &= G(LY, LAGE) + \beta_1 LDRVRS + \beta_2 LSIZE + \beta_3^T Residence \\ &\quad + \beta_4^T Region + \beta_5 Lifecycle + \varepsilon \end{aligned}$$

where LTGALS is log gallons, LY and LAGE denote log(income) and log(age) respectively, LDRVRS is log(numbers of drive), LSIZE is log(household size), and $E(\varepsilon|predictor\ variables) = 0$.

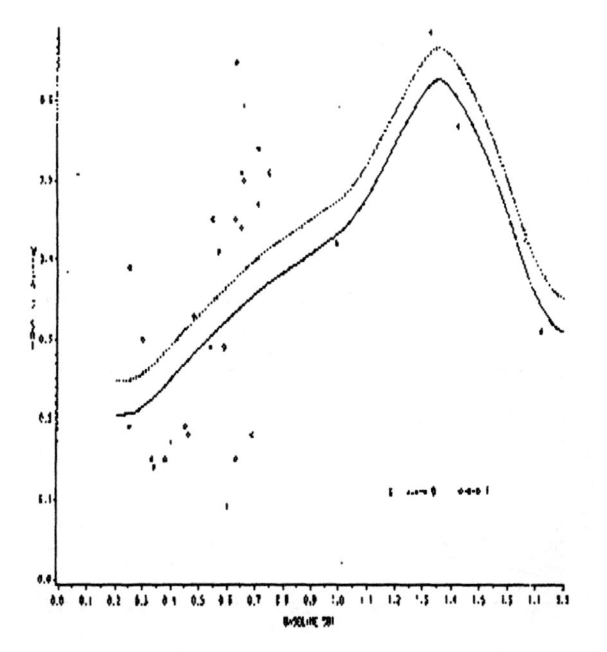

FIGURE 1.2. Raw data partially linear regression estimates for mouthwash data. The predictor variable is $T =$ baseline SBI, the response is $Y = SBI$ index after three weeks. The SBI index is a measurement indicating gum shrinkage. From Speckman (1988), with the permission from the Royal Statistical Society.

Figures 1.3 and 1.4 depicts log-income profiles for different ages and log-age profiles for different incomes. The income structure is quite clear from 1.3. Similarly, 1.4 shows a clear age structure of household gasoline demand.

Example 1.1.4 *Green and Silverman (1994) provided an example of the use of partially linear models, and compared their results with a classical approach employing blocking. They considered the data, primarily discussed by Daniel and Wood (1980), drawn from a marketing price-volume study carried out in the petroleum distribution industry.*

The response variable Y *is the log volume of sales of gasoline, and the two main explanatory variables of interest are* x_1*, the price in cents per gallon of gasoline, and* x_2*, the differential price to competition. The nonparametric component* t *represents the day of the year.*

Their analysis is displayed in Figure 1.5[1]*. Three separate plots against t are*

[1]The postscript files of Figures 1.5-1.7 are provided by Professor Silverman.

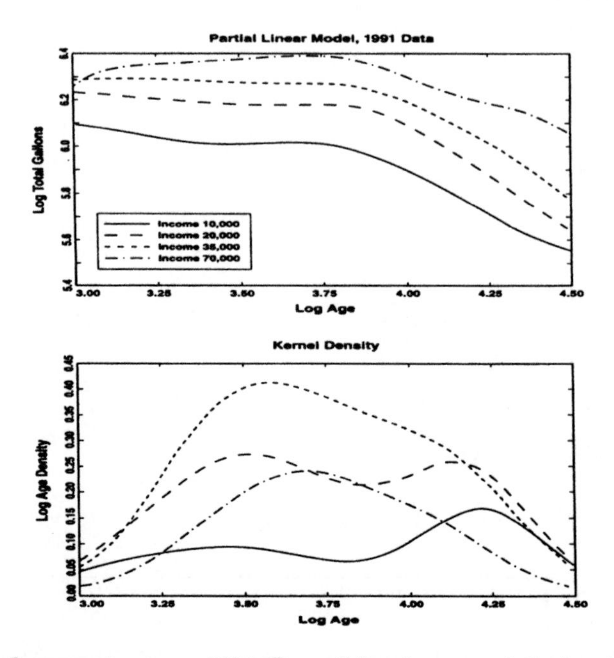

FIGURE 1.3. Income structure, 1991. From Schmalensee and Stoker (1999), with the permission from the Journal of Econometrica.

shown. Upper plot: parametric component of the fit; middle plot: dependence on nonparametric component; lower plot: residuals. All three plots are drown to the same vertical scale, but the upper two plots are displaced upwards.

Example 1.1.5 *Dinse and Lagakos (1983) reported on a logistic analysis of some bioassay data from a US National Toxicology Program study of flame retardants. Data on male and female rates exposed to various doses of a polybrominated biphenyl mixture known as Firemaster FF-1 consist of a binary response variable,* Y*, indicating presence or absence of a particular nonlethal lesion, bile duct hyperplasia, at each animal's death. There are four explanatory variables: log dose,* x_1*, initial weight,* x_2*, cage position (height above the floor),* x_3*, and age at death,* t*. Our choice of this notation reflects the fact that Dinse and Lagakos commented on various possible treatments of this fourth variable. As alternatives to the use of step functions based on age intervals, they considered both a straightforward linear dependence on* t*, and higher order polynomials. In all cases, they fitted a conventional logistic regression model, the GLM data from male and female rats separate in the final analysis, having observed interactions with gender in an*

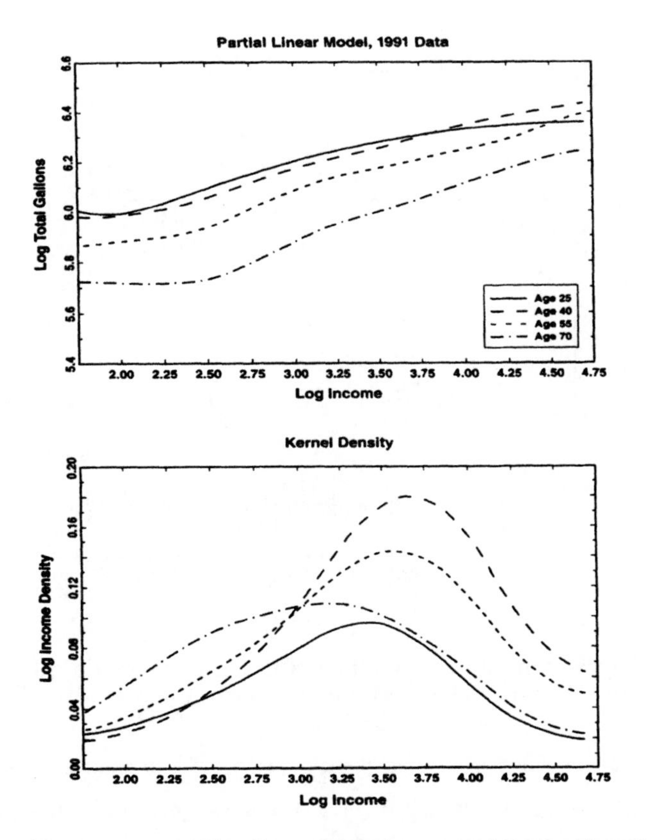

FIGURE 1.4. Age structure, 1991. From Schmalensee and Stoker (1999), with the permission from the Journal of Econometrica.

initial examination of the data.

Green and Yandell (1985) treated this as a semiparametric GLM regression problem, regarding x_1, x_2 and x_3 as linear variables, and t the nonlinear variable. Decompositions of the fitted linear predictors for the male and female rats are shown in Figures 1.6 and 1.7, based on the Dinse and Lagakos data sets, consisting of 207 and 112 animals respectively.

Furthermore, let us now cite two examples of partially linear models that may typically occur in microeconomics, constructed by Tripathi (1997). In these two examples, we are interested in estimating the parametric component when we only know that the unknown function belongs to a set of appropriate functions.

Example 1.1.6 *A firm produces two different goods with production functions F_1 and F_2. That is, $y_1 = F_1(x)$ and $y_2 = F_2(z)$, with $(x \times z) \in R^n \times R^m$. The firm*

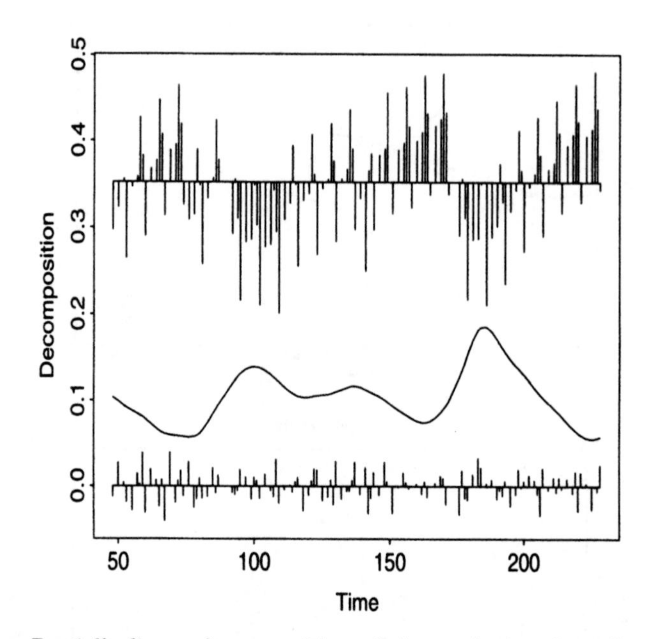

FIGURE 1.5. Partially linear decomposition of the marketing data. Results taken from Green and Silverman (1994) with permission of Chapman & Hall.

maximizes total profits $p_1y_1 - w_1^T x = p_2y_2 - w_2^T z$. *The maximized profit can be written as* $\pi_1(u) + \pi_2(v)$, *where* $u = (p_1, w_1)$ *and* $v = (p_2, w_2)$. *Now suppose that the econometrician has sufficient information about the first good to parameterize the first profit function as* $\pi_1(u) = u^T\theta_0$. *Then the observed profit is* $\pi_i = u_i^T\theta_0 + \pi_2(v_i) + \varepsilon_i$, *where* π_2 *is monotone, convex, linearly homogeneous and continuous in its arguments.*

Example 1.1.7 *Again, suppose we have n similar but geographically dispersed firms with the same profit function. This could happen if, for instance, these firms had access to similar technologies. Now suppose that the observed profit depends not only upon the price vector, but also on a linear index of exogenous variables. That is,* $\pi_i = x_i^T\theta_0 + \pi^*(p_1', \ldots, p_k') + \varepsilon_i$, *where the profit function* π^* *is continuous, monotone, convex, and homogeneous of degree one in its arguments.*

`Partially linear models` are semiparametric models since they contain both parametric and nonparametric components. It allows easier interpretation of the effect of each variable and may be preferred to a completely nonparametric regression because of the well-known "curse of dimensionality". The parametric

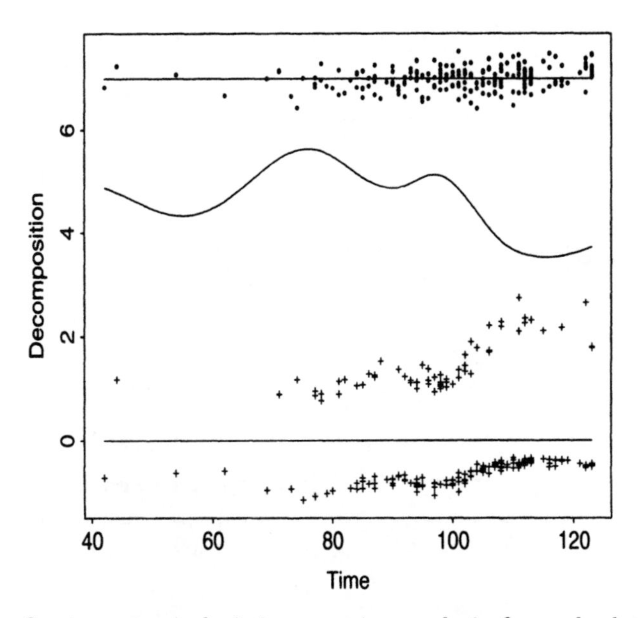

FIGURE 1.6. Semiparametric logistic regression analysis for male data. Results taken from Green and Silverman (1994) with permission of Chapman & Hall.

components can be estimated at the rate of $\sqrt{n}$, while the estimation precision of the nonparametric function decreases rapidly as the the dimension of the nonlinear variable increases. Moreover, the partially linear models are more flexible than the standard linear models, since they combine both parametric and nonparametric components when it is believed that the response depends on some variables in linear relationship but is nonlinearly related to other particular independent variables.

Following the work of Engle, Granger, Rice and Weiss (1986), much attention has been directed to estimating (1.1.1). See, for example, Heckman (1986), Rice (1986), Chen (1988), Robinson (1988), Speckman (1988), Hong (1991), Gao (1992), Liang (1992), Gao and Zhao (1993), Schick (1996a,b) and Bhattacharya and Zhao (1993) and the references therein. For instance, Robinson (1988) constructed a feasible least squares estimator of β based on estimating the nonparametric component by a Nadaraya-Waston kernel estimator. Under some regularity conditions, he deduced the asymptotic distribution of the estimate.

Speckman (1988) argued that the nonparametric component can be charac-

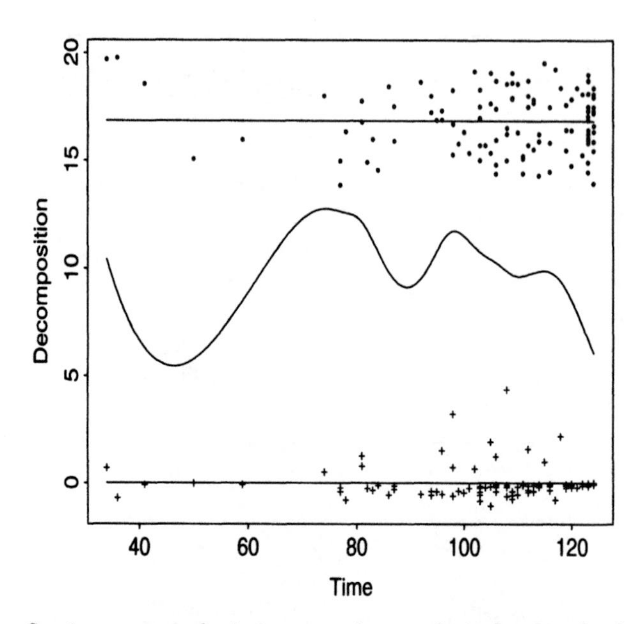

FIGURE 1.7. Semiparametric logistic regression analysis for female data. Results taken from Green and Silverman (1994) with permission of Chapman & Hall.

terized by $\mathcal{W}\gamma$, where $\mathcal{W}$ is a $(n \times q)-$matrix of full rank, γ is an additional unknown parameter and q is unknown. The `partially linear model` (1.1.1) can be rewritten in a matrix form

$$\mathbf{Y} = \mathbf{X}\beta + \mathcal{W}\gamma + \varepsilon. \tag{1.1.2}$$

The estimator of β based on (1.1.2) is

$$\widehat{\beta} = \{\mathbf{X}^T(\mathcal{F} - P_{\mathcal{W}})\mathbf{X})\}^{-1}\{\mathbf{X}^T(\mathcal{F} - P_{\mathcal{W}})\mathbf{Y})\} \tag{1.1.3}$$

where $P_{\mathcal{W}} = \mathcal{W}(\mathcal{W}^T\mathcal{W})^{-1}\mathcal{W}^T$ is a projection matrix. Under some suitable conditions, Speckman (1988) studied the asymptotic behavior of this estimator. This estimator is asymptotically unbiased because β is calculated after removing the influence of T from both the X and Y. (See (3.3a) and (3.3b) of Speckman (1988) and his kernel estimator thereafter). Green, Jennison and Seheult (1985) proposed to replace $\mathcal{W}$ in (1.1.3) by a smoothing operator for estimating β as follows:

$$\widehat{\beta}_{GJS} = \{\mathbf{X}^T(\mathcal{F} - \mathcal{W}_h)\mathbf{X})\}^{-1}\{\mathbf{X}^T(\mathcal{F} - \mathcal{W}_h)\mathbf{Y})\}. \tag{1.1.4}$$

Following Green, Jennison and Seheult (1985), Gao (1992) systematically studied asymptotic behaviors of the least squares estimator given by (1.1.3) for the case of non-random design points.

Engle, Granger, Rice and Weiss (1986), Heckman (1986), Rice (1986), Whaba (1990), Green and Silverman (1994) and Eubank, Kambour, Kim, Klipple, Reese and Schimek (1998) used the spline smoothing technique and defined the penalized estimators of β and g as the solution of

$$\text{argmin}_{\beta,g}\frac{1}{n}\sum_{i=1}^{n}\{Y_i - X_i^T\beta - g(T_i)\}^2 + \lambda\int\{g''(u)\}^2du \tag{1.1.5}$$

where λ is a penalty parameter (see Whaba (1990)). The above estimators are asymptotically biased (Rice, 1986, Schimek, 1997). Schimek (1999) demonstrated in a simulation study that this bias is negligible apart from small sample sizes (e.g. $n = 50$), even when the parametric and nonparametric components are correlated.

The original motivation for Speckman's algorithm was a result of Rice (1986), who showed that within a certain asymptotic framework, the penalized least squares (PLS) estimate of β could be susceptible to biases of the kind that are inevitable when estimating a curve. Heckman (1986) only considered the case where X_i and T_i are independent and constructed an asymptotically normal estimator for β. Indeed, Heckman (1986) proved that the PLS estimator of β is consistent at parametric rates if small values of the smoothing parameter are used. Hamilton and Truong (1997) used local linear regression in partially linear models and established the asymptotic distributions of the estimators of the parametric and nonparametric components. More general theoretical results along with these lines are provided by Cuzick (1992a), who considered the case where the density of ε is known. See also Cuzick (1992b) for an extension to the case where the density function of ε is unknown. Liang (1992) systematically studied the Bahadur efficiency and the `second order asymptotic efficiency` for a numbers of cases. More recently, Golubev and Härdle (1997) derived the upper and lower bounds for the `second minimax order risk` and showed that the `second order minimax estimator` is a penalized maximum likelihood estimator. Similarly, Mammen and van de Geer (1997) applied the theory of empirical processes to derive the asymptotic properties of a penalized `quasi likelihood` estimator, which generalizes the `piecewise polynomial`-based estimator of Chen (1988).

In the case of `heteroscedasticity`, Schick (1996b) constructed root-n consistent weighted least squares estimates and proposed an optimal weight function for the case where the variance function is known up to a multiplicative constant. More recently, Liang and Härdle (1997) further studied this issue for more general variance functions.

Severini and Staniswalis (1994) and Härdle, Mammen and Müller (1998) studied a generalization of (1.1.1), which corresponds to

$$E(Y|X,T) = H\{X^T\beta + g(T)\} \tag{1.1.6}$$

where H (called link function) is a known function, and β and g are the same as in (1.1.1). To estimate β and g, Severini and Staniswalis (1994) introduced the quasi-likelihood estimation method, which has properties similar to those of the likelihood function, but requires only specification of the second-moment properties of Y rather than the entire distribution. Based on the approach of Severini and Staniswalis, Härdle, Mammen and Müller (1998) considered the problem of testing the linearity of g. Their test indicates whether nonlinear shapes observed in nonparametric fits of g are significant. Under the linear case, the test statistic is shown to be asymptotically normal. In some sense, their test complements the work of Severini and Staniswalis (1994). The practical performance of the tests is shown in applications to data on East-West German `migration` and credit scoring. Related discussions can also be found in Mammen and van de Geer (1997) and Carroll, Fan, Gijbels and Wand (1997).

Example 1.1.8 *Consider a model on East–West German migration in 1991 GSOEP (1991)data from the German Socio-Economic Panel for the state Mecklenburg-Vorpommern, a land of the Federal State of Germany. The dependent variable is binary with $Y = 1$ (intention to move) or $Y = 0$ (stay). Let X denote some socioeconomic factors such as age, sex, friends in west, city size and unemployment, T do household income. Figure 1.8 shows a fit of the function g in the semiparametric model (1.1.6). It is clearly nonlinear and shows a saturation in the intention to migrate for higher income households. The question is, of course, whether the observed nonlinearity is significant.*

Example 1.1.9 *Müller and Rönz (2000) discuss credit scoring methods which aim to assess credit worthiness of potential borrowers to keep the risk of credit*

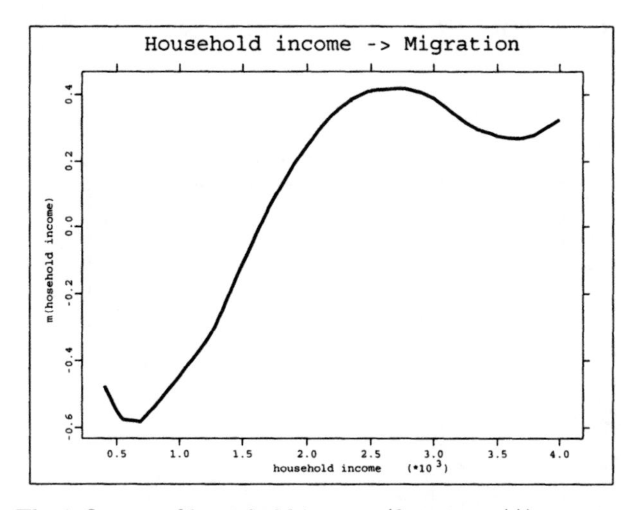

FIGURE 1.8. The influence of household income (function $g(t)$) on migration intention. Sample from Mecklenburg–Vorpommern, $n = 402$.

loss low and to minimize the costs of failure over risk groups. One of the classical parametric approaches, logit regression, assumes that the probability of belonging to the group of "bad" clients is given by $P(Y = 1) = F(\beta^T X)$, *with* $Y = 1$ *indicating a "bad" client and* X *denoting the vector of explanatory variables, which include eight continuous and thirteen categorical variables.* X_2 *to* X_9 *are the continuous variables. All of them have (left) skewed distributions. The variables* X_6 *to* X_9 *in particular have one realization which covers the majority of observations.* X_{10} *to* X_{24} *are the categorical variables. Six of them are dichotomous. The others have 3 to 11 categories which are not ordered. Hence, these variables have been categorized into dummies for the estimation and validation.*

The authors consider a special case of the generalized partially linear model $E(Y|X,T) = G\{\beta^T X + g(T)\}$ *which allows to model the influence of a part* T *of the explanatory variables in a nonparametric way. The model they study is*

$$P(Y = 1) = F\left(g(x_5) + \sum_{j=2, j\neq 5}^{24} \beta_j x_j\right)$$

where a possible constant is contained in the function $g(\cdot)$. *This model is estimated by semiparametric maximum–likelihood, a combination of ordinary and smoothed maximum–likelihood. Figure 1.9 compares the performance of the parametric logit fit and the semiparametric logit fit obtained by including* X_5 *in a nonparametric way. Their analysis indicated that this generalized partially linear model improves the previous performance. The detailed discussion can be found in Müller and Rönz (2000).*

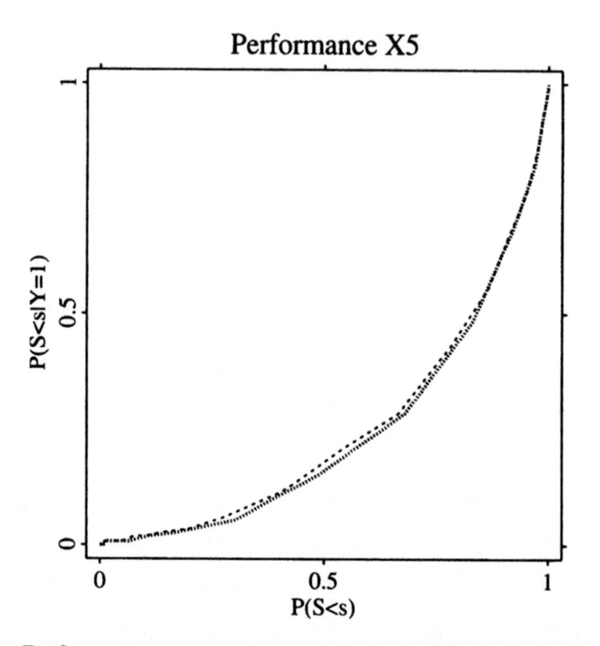

FIGURE 1.9. Performance curves, parametric logit (black dashed) and semiparametric logit (thick grey) with variable X_5 included nonparametrically. Results taken from Müller and Rönz (2000).

1.2 The Least Squares Estimators

If the nonparametric component of the partially linear model is assumed to be known, then LS theory may be applied. In practice, the nonparametric component g, regarded as a nuisance parameter, has to be estimated through smoothing methods. Here we are mainly concerned with the nonparametric regression estimation. For technical convenience, we focus only on the case of $T \in [0,1]$ in Chapters 2-5. In Chapter 6, we extend model (1.1.1) to the multi-dimensional time series case. Therefore some corresponding results for the multidimensional independent case follow immediately, see for example, Sections 6.2 and 6.3.

For identifiability, we assume that the pair (β, g) of (1.1.1) satisfies

$$\frac{1}{n}\sum_{i=1}^{n} E\{Y_i - X_i^T\beta - g(T_i)\}^2 = \min_{(\alpha,f)} \frac{1}{n}\sum_{i=1}^{n} E\{Y_i - X_i^T\alpha - f(T_i)\}^2. \qquad (1.2.1)$$

This implies that if $X_i^T\beta_1+g_1(T_i) = X_i^T\beta_2+g_2(T_i)$ for all $1 \le i \le n$, then $\beta_1 = \beta_2$ and $g_1 = g_2$ simultaneously. We will justify this separately for the random design case and the fixed design case.

For the random design case, if we assume that $E[Y_i|(X_i, T_i)] = X_i^T\beta_1 + g_1(T_i) = X_i^T\beta_2 + g_2(T_i)$ for all $1 \le i \le n$, then it follows from $E\{Y_i - X_i^T\beta_1 - g_1(T_i)\}^2 = E\{Y_i - X_i^T\beta_2 - g_2(T_i)\}^2 + (\beta_1 - \beta_2)^T E\{(X_i - E[X_i|T_i])(X_i - E[X_i|T_i])^T\} (\beta_1 - \beta_2)$ that we have $\beta_1 = \beta_2$ due to the fact that the matrix $E\{(X_i - E[X_i|T_i])(X_i - E[X_i|T_i])^T\}$ is positive definite assumed in Assumption 1.3.1(i) below. Thus $g_1 = g_2$ follows from the fact $g_j(T_i) = E[Y_i|T_i] - E[X_i^T\beta_j|T_i]$ for all $1 \le i \le n$ and $j = 1, 2$.

For the fixed design case, we can justify the identifiability using several different methods. We here provide one of them. Suppose that g of (1.1.1) can be parameterized as $G = \{g(T_1), \ldots, g(T_n)\}^T = W\gamma$ used in (1.2.2), where γ is a vector of unknown parameters.

Then submitting $G = W\gamma$ into (1.2.1), we have the normal equations

$$X^T X\beta = X^T(Y - W\gamma) \quad \text{and} \quad W\gamma = P(Y - X\beta),$$

where $P = W(W^TW)^{-1}W^T$, $X^T = (X_1, \ldots, X_n)$ and $Y^T = (Y_1, \ldots, Y_n)$.

Similarly, if we assume that $E[Y_i] = X_i^T\beta_1 + g_1(T_i) = X_i^T\beta_2 + g_2(T_i)$ for all $1 \le i \le n$, then it follows from Assumption 1.3.1(ii) below and the fact that $1/nE\{(Y - X\beta_1 - W\gamma_1)^T(Y - X\beta_1 - W\gamma_1)\} = 1/nE\{(Y - X\beta_2 - W\gamma_2)^T(Y - X\beta_2 - W\gamma_2)\} + 1/n(\beta_1 - \beta_2)^T X^T(I - P)X(\beta_1 - \beta_2)$ that we have $\beta_1 = \beta_2$ and $g_1 = g_2$ simultaneously.

Assume that $\{(X_i, T_i, Y_i); i = 1, \ldots, n.\}$ satisfies model (1.1.1). Let $\omega_{ni}(t)\{= \omega_{ni}(t; T_1, \ldots, T_n)\}$ be positive weight functions depending on t and the design points $T_1, \ldots, T_n$. For every given β, we define an estimator of $g(\cdot)$ by

$$g_n(t; \beta) = \sum_{i=1}^{n} \omega_{nj}(t)(Y_i - X_i^T\beta).$$

We often drop the β for convenience. Replacing $g(T_i)$ by $g_n(T_i)$ in model (1.1.1) and using the LS criterion, we obtain the least squares estimator of β:

$$\beta_{LS} = (\widetilde{\mathbf{X}}^T\widetilde{\mathbf{X}})^{-1}\widetilde{\mathbf{X}}^T\widetilde{\mathbf{Y}}, \tag{1.2.2}$$

which is just the estimator $\widehat{\beta}_{GJS}$ in (1.1.4) with a different smoothing operator.

The nonparametric estimator of $g(t)$ is then defined as follows:

$$\widehat{g}_n(t) = \sum_{i=1}^{n} \omega_{ni}(t)(Y_i - X_i^T \beta_{LS}). \tag{1.2.3}$$

where $\widetilde{\mathbf{X}}^T = (\widetilde{X}_1, \ldots, \widetilde{X}_n)$ with $\widetilde{X}_j = X_j - \sum_{i=1}^n \omega_{ni}(T_j)X_i$ and $\widetilde{\mathbf{Y}}^T = (\widetilde{Y}_1, \ldots, \widetilde{Y}_n)$ with $\widetilde{Y}_j = Y_j - \sum_{i=1}^n \omega_{ni}(T_j)Y_i$. Due to Lemma A.2 below, we have as $n \to \infty$ $n^{-1}(\widetilde{\mathbf{X}}^T\widetilde{\mathbf{X}}) \to \Sigma$, where Σ is a positive matrix. Thus, we assume that $n(\widetilde{\mathbf{X}}^T\widetilde{\mathbf{X}})^{-1}$ exists for large enough n throughout this monograph.

When $\varepsilon_1, \ldots, \varepsilon_n$ are identically distributed, we denote their distribution function by $\varphi(\cdot)$ and the variance by σ^2, and define the estimator of σ^2 by

$$\widehat{\sigma}_n^2 = \frac{1}{n} \sum_{i=1}^{n} (\widetilde{Y}_i - \widetilde{X}_i^T \beta_{LS})^2 \tag{1.2.4}$$

In this monograph, most of the estimation procedures are based on the estimators (1.2.2), (1.2.3) and (1.2.4).

1.3 Assumptions and Remarks

This monograph considers the two cases: the `fixed design` and the i.i.d. `random design`. When considering the random case, denote

$$h_j(T_i) = E(x_{ij}|T_i) \text{ and } u_{ij} = x_{ij} - E(x_{ij}|T_i).$$

Assumption 1.3.1 *i)* $\sup_{0 \le t \le 1} E(\|X_1\|^3 | T = t) < \infty$ *and* $\Sigma = Cov\{X_1 - E(X_1|T_1)\}$ *is a positive definite matrix. The random errors* ε_i *are independent of* (X_i, T_i).

ii) When (X_i, T_i) *are* `fixed design points`, *there exist continuous functions* $h_j(\cdot)$ *defined on* $[0, 1]$ *such that each component of* X_i *satisfies*

$$x_{ij} = h_j(T_i) + u_{ij} \quad 1 \le i \le n, \quad 1 \le j \le p \tag{1.3.1}$$

where $\{u_{ij}\}$ *is a sequence of real numbers satisfying*

$$\lim_{n\to\infty} \frac{1}{n} \sum_{i=1}^{n} u_i u_i^T = \Sigma \tag{1.3.2}$$

and for $m = 1, \ldots, p$,

$$\limsup_{n\to\infty} \frac{1}{a_n} \max_{1 \le k \le n} \left| \sum_{i=1}^{k} u_{j_i m} \right| < \infty \tag{1.3.3}$$

for all permutations $(j_1, \ldots, j_n)$ *of* $(1, 2, \ldots, n)$, *where* $u_i = (u_{i1}, \ldots, u_{ip})^T$, $a_n = n^{1/2} \log n$, *and* Σ *is a positive definite matrix.*

Throughout the monograph, we apply Assumption 1.3.1 i) to the case of **random design points** and Assumption 1.3.1 ii) to the case where (X_i, T_i) are **fixed design points**. Assumption 1.3.1 i) is a reasonable condition for the random design case, while Assumption 1.3.1 ii) generalizes the corresponding conditions of Heckman (1986) and Rice (1986), and simplifies the conditions of Speckman (1988). See also Remark 2.1 (i) of Gao and Liang (1997).

Assumption 1.3.2 *The first two derivatives of $g(\cdot)$ and $h_j(\cdot)$ are* **Lipschitz continuous** *of order one.*

Assumption 1.3.3 *When (X_i, T_i) are fixed design points, the positive weight functions $\omega_{ni}(\cdot)$ satisfy*

$$
\begin{aligned}
(i) \quad & \max_{1\le i\le n} \sum_{j=1}^{n} \omega_{ni}(T_j) = O(1), \\
& \max_{1\le j\le n} \sum_{i=1}^{n} \omega_{ni}(T_j) = O(1), \\
(ii) \quad & \max_{1\le i,j\le n} \omega_{ni}(T_j) = O(b_n), \\
(iii) \quad & \max_{1\le i\le n} \sum_{j=1}^{n} \omega_{nj}(T_i) I(|T_i - T_j| > c_n) = O(c_n),
\end{aligned}
$$

where b_n and c_n are two sequences satisfying $\limsup_{n\to\infty} nb_n^2 \log^4 n < \infty$, $\liminf_{n\to\infty} nc_n^2 > 0$, $\limsup_{n\to\infty} nc_n^4 \log n < \infty$ and $\limsup_{n\to\infty} nb_n^2 c_n^2 < \infty$. When (X_i, T_i) are i.i.d. random design points, (i), (ii) and (iii) hold with probability one.

Remark 1.3.1 *There are many weight functions satisfying Assumption 1.3.3. For examples,*

$$
W_{ni}^{(1)}(t) = \frac{1}{h_n}\int_{S_{i-1}}^{S_i} K\Big(\frac{t-s}{H_n}\Big)ds, \quad W_{ni}^{(2)}(t) = K\Big(\frac{t-T_i}{H_n}\Big) \Big/ \sum_{j=1}^{n} K\Big(\frac{t-T_j}{H_n}\Big),
$$

where $S_i = \frac{1}{2}(T_{(i)} + T_{(i-1)})$, $i = 1, \cdots, n-1$, $S_0 = 0$, $S_n = 1$, and $T_{(i)}$ are the order statistics of $\{T_i\}$. $K(\cdot)$ is a kernel function satisfying certain conditions, and H_n is a positive number sequence. Here $H_n = h_n$ or r_n, h_n is a **bandwidth** *parameter, and $r_n = r_n(t, T_1, \cdots, T_n)$ is the distance from t to the k_n-th nearest neighbor among the T_i's, and where k_n is an integer sequence.*

We can justify that both $W_{ni}^{(1)}(t)$ and $W_{ni}^{(2)}(t)$ satisfy Assumption 1.3.3. The details of the justification are very lengthy and omitted. We also want to point out that when ω_{ni} is either $W_{ni}^{(1)}$ or $W_{ni}^{(2)}$, Assumption 1.3.3 holds automatically with $H_n = \lambda n^{-1/5}$ for some $0 < \lambda < \infty$. This is the same as the result established by Speckman (1988) (see Theorem 2 with $\nu = 2$), who pointed out that the usual $n^{-1/5}$ rate for the bandwidth is fast enough to establish that the LS estimate β_{LS} of β is $\sqrt{n}$-consistent. Sections 2.1.3 and 6.4 will discuss some practical selections for the bandwidth.

Remark 1.3.2 *Throughout this monograph, we are mostly using Assumption 1.3.1 ii) and 1.3.3 for the fixed design case. As a matter of fact, we can replace Assumption 1.3.1 ii) and 1.3.3 by the following corresponding conditions.*

Assumption 1.3.1 *ii)' When (X_i, T_i) are the fixed design points, equations (1.3.1) and (1.3.2) hold.*

Assumption 1.3.3' *When (X_i, T_i) are fixed design points, Assumption 1.3.3 (i)-(iii) holds. In addition, the weight functions ω_{ni} satisfy*

$$(iv) \quad \max_{1\le i\le n} \sum_{j=1}^{n} \omega_{nj}(T_i)u_{jl} = O(d_n),$$

$$(v) \quad \frac{1}{n}\sum_{j=1}^{n} \widetilde{f}_j u_{jl} = O(d_n),$$

$$(vi) \quad \frac{1}{n}\sum_{j=1}^{n}\Big\{\sum_{k=1}^{n} \omega_{nk}(T_j)u_{ks}\Big\}u_{jl} = O(d_n)$$

for all $1 \le l, s \le p$, where d_n is a sequence of real numbers satisfying $\limsup_{n\to\infty} nd_n^4 \log n < \infty$, $\widetilde{f}_j = f(T_j) - \sum_{k=1}^{n} \omega_{nk}(T_j)f(T_k)$ for $f = g$ or h_j defined in (1.3.1).

Obviously, the three conditions (iv), (v) and (vi) follows from (1.3.3) and Abel's inequality.

When the weight functions ω_{ni} are chosen as $W_{ni}^{(2)}$ defined in Remark 1.3.1, Assumptions 1.3.1 ii)' and 1.3.3' are almost the same as Assumptions (a)-(f) of Speckman (1988). As mentioned above, however, we prefer to use Assumptions 1.3.1 ii) and 1.3.3 for the fixed design case throughout this monograph.

Under the above assumptions, we provide bounds for $h_j(T_i) - \sum_{k=1}^{n} \omega_{nk}(T_i) h_j(T_k)$ and $g(T_i) - \sum_{k=1}^{n} \omega_{nk}(T_i)g(T_k)$ in the appendix.

1.4 The Scope of the Monograph

The main objectives of this monograph are: (i) To present a number of theoretical results for the estimators of both parametric and nonparametric components, and (ii) To illustrate the proposed estimation and testing procedures by several simulated and true data sets using `XploRe`-The Interactive Statistical Computing Environment (see Härdle, Klinke and Müller, 1999), available on website: `http://www.xplore-stat.de`.

In addition, we generalize the existing approaches for **homoscedasticity** to **heteroscedastic** models, introduce and study partially linear errors-in-variables models, and discuss **partially linear time series** models.

1.5 The Structure of the Monograph

The monograph is organized as follows: Chapter 2 considers a simple partially linear model. An estimation procedure for the parametric component of the partially linear model is established based on the nonparametric weight sum. Section 2.1 mainly provides asymptotic theory and an estimation procedure for the parametric component with **heteroscedastic** errors. In this section, the least squares estimator β_{LS} of (1.2.2) is modified to the weighted least squares estimator β_{WLS}. For constructing β_{WLS}, we employ the **split-sample** techniques. The asymptotic normality of β_{WLS} is then derived. Three different variance functions are discussed and estimated. The selection of smoothing parameters involved in the nonparametric weight sum is also discussed in Subsection 2.1.3. Simulation comparison is also implemented in Subsection 2.1.4. A modified estimation procedure for the case of **censored** data is given in Section 2.2. Based on a modification of the **Kaplan-Meier estimator**, **synthetic data** and an estimator of β are constructed. We then establish the asymptotic normality for the resulting estimator of β. We also examine the behaviors of the finite sample through a simulated example. Bootstrap approximations are given in Section 2.3.

Chapter 3 discusses the estimation of the nonparametric component without the restriction of constant variance. Convergence and asymptotic normality of the nonparametric estimate are given in Sections 3.2 and 3.3. The estimation methods proposed in this chapter are illustrated through examples in Section 3.4, in which the estimator (1.2.3) is applied to the analysis of the logarithm of the earnings to labour market experience.

In Chapter 4, we consider both linear and nonlinear variables with measurement errors. An estimation procedure and asymptotic theory for the case where the linear variables are measured with measurement errors are given in Section 4.1. The common estimator given in (1.2.2) is modified by applying the so-called "`correction for attenuation`", and hence deletes the inconsistence caused by `measurement error`. The modified estimator is still asymptotically normal as (1.2.2) but with a more complicated form of the asymptotic variance. Section 4.2 discusses the case where the nonlinear variables are measured with measurement errors. Our conclusion shows that asymptotic normality heavily depends on the distribution of the `measurement error` when T is measured with error. Examples and numerical discussions are presented to support the theoretical results.

Chapter 5 discusses several relatively theoretic topics. The `laws of the iterative logarithm` (LIL) and the Berry-Esseen bounds for the parametric component are established. Section 5.3 constructs a class of `asymptotically efficient` estimators of β. Two classes of efficiency concepts are introduced. The well-known Bahadur asymptotic efficiency, which considers the exponential rate of the `tail probability`, and second order asymptotic efficiency are discussed in detail in Sections 5.4 and 5.5, respectively. The results of this chapter show that the LS estimate can be modified to have both Bahadur asymptotic efficiency and second order asymptotic efficiency even when the parametric and nonparametric components are dependent. The estimation of the error distribution is also investigated in Section 5.6.

Chapter 6 generalizes the case studied in previous chapters to `partially linear time series` models and establishes asymptotic results as well as small sample studies. At first we present several data-based `test statistics` to determine which model should be chosen to model a partially linear dynamical system. Secondly we propose a cross-validation (CV) based criterion to select the optimum linear subset for a partially linear regression model. We investigate the problem of selecting the optimum bandwidth for a `partially linear autoregressive model`. Finally, we summarize recent developments in a general class of `additive stochastic regression models`.

2
ESTIMATION OF THE PARAMETRIC COMPONENT

2.1 Estimation with Heteroscedastic Errors

2.1.1 Introduction

This section considers asymptotic normality for the estimator of β when ε is a `homoscedastic` error. This aspect has been discussed by Chen (1988), Robinson (1988), Speckman (1988), Hong (1991), Gao and Zhao (1993), Gao, Chen and Zhao (1994) and Gao, Hong and Liang (1995). Here we state one of the main results obtained by Gao, Hong and Liang (1995) for model (1.1.1).

Theorem 2.1.1 *Under Assumptions 1.3.1-1.3.3, β_{LS} is an asymptotically normal estimator of β, i.e.,*

$$\sqrt{n}(\beta_{LS} - \beta) \longrightarrow^{\mathcal{L}} N(0, \sigma^2 \Sigma^{-1}). \tag{2.1.1}$$

Furthermore, assume that the weight functions $\omega_{ni}(t)$ are `Lipschitz continuous` *of order one. Let $\sup_i E|\varepsilon_i|^3 < \infty$, $b_n = n^{-4/5} \log^{-1/5} n$ and $c_n = n^{-2/5} \log^{2/5} n$ in Assumption 1.3.3. Then with probability one*

$$\sup_{0 \le t \le 1} |\widehat{g}_n(t) - g(t)| = O(n^{-2/5} \log^{2/5} n). \tag{2.1.2}$$

The proof of this theorem has been given in several papers. The proof of (2.1.1) is similar to that of Theorem 2.1.2 below. Similar to the proof of Theorem 5.1 of Müller and Stadtmüller (1987), the proof of (2.1.2) can be completed. The details have been given in Gao, Hong and Liang (1995).

Example 2.1.1 *Suppose the data are drawn from $Y_i = X_i^T \beta_0 + T_i^3 + \varepsilon_i$ for $i = 1, \ldots, 100$, where $\beta_0 = (1.2, 1.3, 1.4)^T$, $T_i \sim U[0,1]$, $\varepsilon_i \sim N(0, 0.01)$ and $X_i \sim N(0, \Sigma_x)$ with $\Sigma_x = \begin{pmatrix} 0.81 & 0.1 & 0.2 \\ 0.1 & 2.25 & 0.1 \\ 0.2 & 0.1 & 1 \end{pmatrix}$. In this simulation, we perform 20 replications and take bandwidth 0.05. The estimate β_{LS} is $(1.201167, 1.30077, 1.39774)^T$ with mean squared error $(2.1 * 10^{-5}, 2.23 * 10^{-5}, 5.1 * 10^{-5})^T$. The estimate of $g_0(t) (= t^3)$ is based on (1.2.3). For comparison, we also calculate a parametric* *fit for $g_0(t)$. Figure 2.1 shows the parametric estimate and nonparametric fitting for $g_0(t)$. The true curve is given by grey line(in the left side), the nonparametric estimate by thick curve(in the right side) and the parametric estimate by the black straight line.*

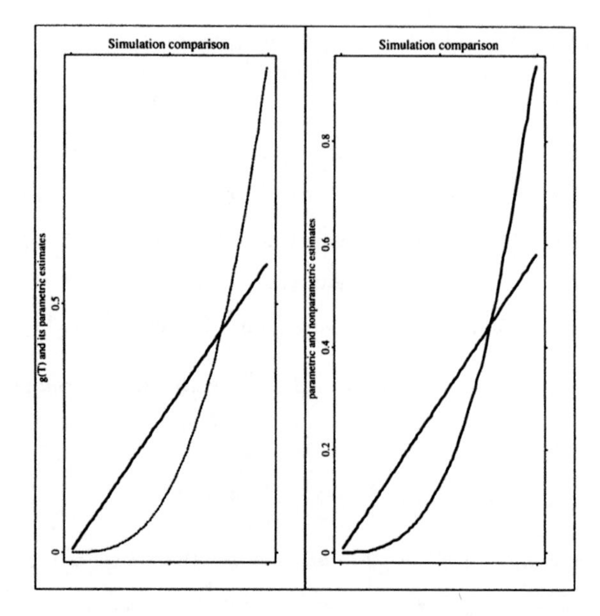

FIGURE 2.1. Parametric and nonparametric estimates of the function $g(T)$

Schick (1996b) considered the problem of **heteroscedasticity**, i.e., nonconstant variance, for model (1.1.1). He constructed root-n consistent **weighted least squares estimates** for the case where the variance is known up to a multiplicative constant. In his discussion, he assumed that the nonconstant variance function of Y given (X, T) is an unknown smooth function of an **exogenous** random vector W.

In the remainder of this section, we mainly consider model (1.1.1) with **heteroscedastic** error and focus on the following cases: (i) $\{\sigma_i^2\}$ is an unknown function of independent **exogenous** variables; (ii) $\{\sigma_i^2\}$ is an unknown function of T_i; and (iii) $\{\sigma_i^2\}$ is an unknown function of $X_i^T\beta + g(T_i)$. We establish asymptotic results for the three cases. In relation to our results, we mention recent developments in linear and nonparametric regression models with **heteroscedastic** errors. See for example, Bickel (1978), Box and Hill (1974), Carroll (1982), Carroll and Ruppert (1982), Carroll and Härdle (1989), Fuller and Rao (1978), Hall and Carroll (1989), Jobson and Fuller (1980), Mak (1992) and Müller and Stadtmüller (1987).

Let $\{(Y_i, X_i, T_i), i = 1, \ldots, n\}$ denote a sequence of random samples from

$$Y_i = X_i^T \beta + g(T_i) + \sigma_i \xi_i, i = 1, \ldots, n, \tag{2.1.3}$$

where (X_i, T_i) are i.i.d. random variables, ξ_i are i.i.d. with mean 0 and variance 1, and σ_i^2 are some functions of other variables. The concrete forms of σ_i^2 will be discussed in later subsections.

When the errors are **heteroscedastic**, β_{LS} is modified to a **weighted least squares** estimator

$$\beta_W = \Big(\sum_{i=1}^n \gamma_i \widetilde{X}_i \widetilde{X}_i^T\Big)^{-1} \Big(\sum_{i=1}^n \gamma_i \widetilde{X}_i \widetilde{Y}_i\Big) \tag{2.1.4}$$

for some weights γ_i $i = 1, \ldots, n$. In this section, we assume that $\{\gamma_i\}$ is either a sequence of random variables or a sequence of constants. In our model (2.1.3) we take $\gamma_i = 1/\sigma_i^2$.

In principle the weights γ_i (or σ_i^2) are unknown and must be estimated. Let $\{\widehat{\gamma}_i, i = 1, \ldots, n\}$ be a sequence of estimators of $\{\gamma_i\}$. We define an estimator of β by substituting γ_i in (2.1.4) by $\widehat{\gamma}_i$.

In order to develop the asymptotic theory conveniently, we use the technique of **split-sample**. Let $k_n (\le n/2)$ be the largest integer part of $n/2$. $\widehat{\gamma}_i^{(1)}$ and $\widehat{\gamma}_i^{(2)}$ are the estimators of γ_i based on the first k_n observations (X_1, T_1, Y_1), $\ldots$, $(X_{k_n}, T_{k_n}, Y_{k_n})$, and the later $n - k_n$ observations $(X_{k_n+1}, T_{k_n+1}, Y_{k_n+1})$, $\ldots$, (X_n, T_n, Y_n), respectively. Define

$$\beta_{WLS} = \Big(\sum_{i=1}^n \widehat{\gamma}_i \widetilde{X}_i \widetilde{X}_i^T\Big)^{-1} \Big(\sum_{i=1}^{k_n} \widehat{\gamma}_i^{(2)} \widetilde{X}_i \widetilde{Y}_i + \sum_{i=k_n+1}^{n} \widehat{\gamma}_i^{(1)} \widetilde{X}_i \widetilde{Y}_i\Big) \tag{2.1.5}$$

as the estimator of β.

The next step is to prove that β_{WLS} is asymptotically normal. We first prove that β_W is asymptotically normal, and then show that $\sqrt{n}(\beta_{WLS} - \beta_W)$ converges to zero in probability.

Assumption 2.1.1 $\sup_{0 \le t \le 1} E(\|X_1\|^3 | T = t) < \infty$. *When* $\{\gamma_i\}$ *is a sequence of real numbers, then* $\lim_{n\to\infty} 1/n \sum_{i=1}^n \gamma_i u_i u_i^T = B$, *where* B *is a positive definite matrix, and* $\lim_{n\to\infty} 1/n \sum_{i=1}^n \gamma_i < \infty$. *When* $\{\gamma_i\}$ *is a sequence of i.i.d. random variables, then* $B = E(\gamma_1 u_1 u_1^T)$ *is a positive definite matrix.*

Assumption 2.1.2 *There exist constants C_1 and C_2 such that*

$$0 < C_1 \le \min_{i \le n} \gamma_i \le \max_{i \le n} \gamma_i < C_2 < \infty.$$

We suppose that the estimator $\{\widehat{\gamma}_i\}$ of $\{\gamma_i\}$ satisfy

$$\max_{1 \le i \le n} |\widehat{\gamma}_i - \gamma_i| = o_P(n^{-q}) \quad q \ge 1/4. \tag{2.1.6}$$

We shall construct estimators to satisfy (2.1.6) for three kinds of γ_i later. The following theorems present general results for the estimators of the parametric components in the partially linear heteroscedastic model (2.1.3).

Theorem 2.1.2 *Assume that Assumptions 2.1.1, 2.1.2 and 1.3.2-1.3.3 hold. Then β_W is an asymptotically normal estimator of β, i.e.,*

$$\sqrt{n}(\beta_W - \beta) \longrightarrow^{\mathcal{L}} N(0, B^{-1}\Sigma B^{-1}).$$

Theorem 2.1.3 *Under Assumptions 2.1.1, 2.1.2 and (2.1.6), β_{WLS} is asymptotically equivalent to β_W, i.e., $\sqrt{n}(\beta_{WLS} - \beta)$ and $\sqrt{n}(\beta_W - \beta)$ have the same asymptotically normal distribution.*

Remark 2.1.1 *In the case of constant error variance, i.e. $\sigma_i^2 \equiv \sigma^2$, Theorem 2.1.2 has been obtained by many authors. See, for example, Theorem 2.1.1.*

Remark 2.1.2 *Theorem 2.1.3 not only assures that our estimator given in (2.1.5) is asymptotically equivalent to the weighted LS estimator with known weights, but also generalizes the results obtained previously.*

Before proving Theorems 2.1.2 and 2.1.3, we discuss three different variance functions and construct their corresponding estimates. Subsection 2.1.4 gives small sample simulation results. The proofs of Theorems 2.1.2 and 2.1.3 are postponed to Subsection 2.1.5.

2.1.2 Estimation of the Non-constant Variance Functions

2.1.2.1 Variance is a Function of Exogenous Variables

This subsection is devoted to the nonparametric heteroscedasticity structure

$$\sigma_i^2 = H(W_i),$$

where H is unknown and **Lipschitz continuous**, $\{W_i; i = 1, \ldots, n\}$ is a sequence of i.i.d. design points defined on $[0, 1]$, which are assumed to be independent of (ξ_i, X_i, T_i).

Define

$$\widehat{H}_n(w) = \sum_{j=1}^{n} \widetilde{\omega}_{nj}(w)\{Y_j - X_j^T \beta_{LS} - \widehat{g}_n(T_i)\}^2$$

as the estimator of $H(w)$, where $\{\widetilde{\omega}_{nj}(t); j = 1, \ldots, n\}$ is a sequence of weight functions satisfying Assumption 1.3.3 with ω_{nj} replaced by $\widetilde{\omega}_{nj}$.

Theorem 2.1.4 *Assume that the conditions of Theorem 2.1.2 hold. Let $c_n = n^{-1/3}\log n$ in Assumption 1.3.3. Then*

$$\sup_{1 \le i \le n} |\widehat{H}_n(W_i) - H(W_i)| = O_P(n^{-1/3}\log n).$$

Proof. Note that

$$\begin{aligned}
\widehat{H}_n(W_i) &= \sum_{j=1}^{n} \widetilde{\omega}_{nj}(W_i)(\widetilde{Y}_j - \widetilde{X}_j^T \beta_{LS})^2 \\
&= \sum_{j=1}^{n} \widetilde{\omega}_{nj}(W_i)\{\widetilde{X}_j^T(\beta - \beta_{LS}) + \widetilde{g}(T_i) + \widetilde{\varepsilon}_i\}^2 \\
&= (\beta - \beta_{LS})^T \sum_{j=1}^{n} \widetilde{\omega}_{nj}(W_i)\widetilde{X}_j\widetilde{X}_j^T(\beta - \beta_{LS}) + \sum_{j=1}^{n} \widetilde{\omega}_{nj}(W_i)\widetilde{g}^2(T_i) \\
&\quad + \sum_{j=1}^{n} \widetilde{\omega}_{nj}(W_i)\widetilde{\varepsilon}_i^2 + 2\sum_{j=1}^{n} \widetilde{\omega}_{nj}(W_i)\widetilde{X}_j^T(\beta - \beta_{LS})\widetilde{g}(T_i) \\
&\quad +2\sum_{j=1}^{n} \widetilde{\omega}_{nj}(W_i)\widetilde{X}_j^T(\beta - \beta_{LS})\widetilde{\varepsilon}_i + 2\sum_{j=1}^{n} \widetilde{\omega}_{nj}(W_i)\widetilde{g}(T_i)\widetilde{\varepsilon}_i. \qquad (2.1.7)
\end{aligned}$$

The first term of (2.1.7) is therefore $O_P(n^{-2/3})$ since $\sum_{j=1}^{n} \widetilde{X}_j\widetilde{X}_j^T$ is a symmetric matrix, $0 < \widetilde{\omega}_{nj}(W_i) \le Cn^{-2/3}$,

$$\sum_{j=1}^{n}\{\widetilde{\omega}_{nj}(W_i) - Cn^{-2/3}\}\widetilde{X}_j\widetilde{X}_j^T$$

is a $p \times p$ nonpositive matrix, and $\beta_{LS} - \beta = O_P(n^{-1/2})$. The second term of (2.1.7) is easily shown to be of order $O_P(n^{1/3}c_n^2)$.

Now we need to prove

$$\sup_i \Big|\sum_{j=1}^{n} \widetilde{\omega}_{nj}(W_i)\widetilde{\varepsilon}_i^2 - H(W_i)\Big| = O_P(n^{-1/3}\log n), \qquad (2.1.8)$$

which is equivalent to proving the following three results

$$\sup_i \Big|\sum_{j=1}^n \widetilde{\omega}_{nj}(W_i)\Big\{\sum_{k=1}^n \omega_{nk}(T_j)\varepsilon_k\Big\}^2\Big| = O_P(n^{-1/3}\log n), \tag{2.1.9}$$

$$\sup_i \Big|\sum_{j=1}^n \widetilde{\omega}_{nj}(W_i)\varepsilon_i^2 - H(W_i)\Big| = O_P(n^{-1/3}\log n), \tag{2.1.10}$$

$$\sup_i \Big|\sum_{j=1}^n \widetilde{\omega}_{nj}(W_i)\varepsilon_j\Big\{\sum_{k=1}^n \omega_{nk}(T_j)\varepsilon_k\Big\}\Big| = O_P(n^{-1/3}\log n). \tag{2.1.11}$$

(A.3) below assures that (2.1.9) holds. Lipschitz continuity of $H(\cdot)$ and assumptions on $\widetilde{\omega}_{nj}(\cdot)$ imply that

$$\sup_i \Big|\sum_{j=1}^n \widetilde{\omega}_{nj}(W_i)\varepsilon_i^2 - H(W_i)\Big| = O_P(n^{-1/3}\log n). \tag{2.1.12}$$

By taking $a_{ki} = \widetilde{\omega}_{nk}(W_i)H(W_k)$, $V_k = \xi_k^2 - 1$ $r = 2$, $p_1 = 2/3$ and $p_2 = 0$ in Lemma A.3, we have

$$\sup_i \Big|\sum_{j=1}^n \widetilde{\omega}_{nj}(W_i)H(W_j)(\xi_j^2 - 1)\Big| = O_P(n^{-1/3}\log n). \tag{2.1.13}$$

A combination of (2.1.13) and (2.1.12) implies (2.1.10). Cauchy-Schwarz inequality, (2.1.9) and (2.1.10) imply (2.1.11), and then (2.1.8). The last three terms of (2.1.7) are all of order $O_P(n^{-1/3}\log n)$ by Cauchy-Schwarz inequality. We therefore complete the proof of Theorem 2.1.4.

2.1.2.2 Variance is a Function of the Design Points T_i

In this subsection we consider the case where $\{\sigma_i^2\}$ is a function of $\{T_i\}$, i.e.,

$$\sigma_i^2 = H(T_i), \ \ H \text{ unknown Lipschitz continuous.}$$

Similar to Subsection 2.1.2.1, we define our estimator of $H(\cdot)$ as

$$\widehat{H}_n(t) = \sum_{j=1}^n \widetilde{\omega}_{nj}(t)\{Y_j - X_j^T\beta_{LS} - \widehat{g}_n(T_i)\}^2.$$

Theorem 2.1.5 *Under the conditions of Theorem 2.1.2, we have*

$$\sup_{1\le i\le n} |\widehat{H}_n(T_i) - H(T_i)| = O_P(n^{-1/3}\log n).$$

Proof. The proof of Theorem 2.1.5 is similar to that of Theorem 2.1.4 and therefore omitted.

2.1.2.3 Variance is a Function of the Mean

Here we consider model (2.1.3) with

$$\sigma_i^2 = H\{X_i^T\beta + g(T_i)\}, \text{ } H \text{ unknown Lipschitz continuous.}$$

This means that the variance is an unknown function of the mean response. Several related situations in linear and nonlinear models have been discussed by Box and Hill (1974), Bickel (1978), Jobson and Fuller (1980), Carroll (1982) and Carroll and Ruppert (1982).

Since $H(\cdot)$ is assumed to be completely unknown, the standard method is to get information about $H(\cdot)$ by replication, i.e., to consider the following "improved" partially linear heteroscedastic model

$$Y_{ij} = X_i^T\beta + g(T_i) + \sigma_i\xi_{ij}, \quad j = 1, \ldots, m_i; \quad i = 1, \ldots, n,$$

where $\{Y_{ij}\}$ is the response of the $j-$th replicate at the design point (X_i, T_i), ξ_{ij} are i.i.d. with mean 0 and variance 1, β, $g(\cdot)$ and (X_i, T_i) are as defined in (2.1.3).

We here apply the idea of Fuller and Rao (1978) for linear heteroscedastic model to construct an estimate of σ_i^2. Based on the least squares estimate β_{LS} and the nonparametric estimate $\widehat{g}_n(T_i)$, we use $Y_{ij} - \{X_i^T\beta_{LS} + \widehat{g}_n(T_i)\}$ to define

$$\widehat{\sigma}_i^2 = \frac{1}{m_i}\sum_{j=1}^{m_i}[Y_{ij} - \{X_i^T\beta_{LS} + \widehat{g}_n(T_i)\}]^2, \tag{2.1.14}$$

with a positive sequence $\{m_i; \, i = 1, \cdots, n\}$ determined later.

Theorem 2.1.6 *Let* $m_i = a_n n^{2q} \stackrel{\text{def}}{=} m(n)$ *for some sequence* a_n *converging to infinity. Suppose the conditions of Theorem 2.1.2 hold. Then*

$$\sup_{1\le i\le n} |\widehat{\sigma}_i^2 - H\{X_i^T\beta + g(T_i)\}| = o_P(n^{-q}) \quad q \ge 1/4.$$

Proof. We provide only an outline for the proof of Theorem 2.1.6. Obviously

$$\begin{aligned}|\widehat{\sigma}_i^2 - H\{X_i^T\beta + g(T_i)\}| \le 3\{X_i^T(\beta - \beta_{LS})\}^2 &+ 3\{g(T_i) - \widehat{g}_n(T_i)\}^2 \\ &+ \frac{3}{m_i}\sum_{j=1}^{m_i}\sigma_i^2(\xi_{ij}^2 - 1).\end{aligned}$$

The first two items are obviously $o_P(n^{-q})$. Since ξ_{ij} are i.i.d. with mean zero and variance 1, by taking $m_i = a_n n^{2q}$, using the law of the iterated logarithm and the boundedness of $H(\cdot)$, we have

$$\frac{1}{m_i}\sum_{j=1}^{m_i}\sigma_i^2(\xi_{ij}^2 - 1) = O\{m(n)^{-1/2}\log m(n)\} = o_P(n^{-q}).$$

Thus we derive the proof of Theorem 2.1.6.

2.1.3 Selection of Smoothing Parameters

In practice, an important problem is how to select the smoothing parameter involved in the weight functions ω_{ni}. Currently, the results on bandwidth selection for completely nonparametric regression can be found in the monographs by Eubank (1988), Härdle (1990, 1991), Wand and Jones (1994), Fan and Gijbels (1996), and Bowman and Azzalini (1997).

More recently, Gao and Anh (1999) considered the selection of an optimum truncation parameter for model (1.1.1) and established large and small sample results for the case where the weight functions ω_{ni} are a sequence of orthogonal series. See also Gao (1998), who discussed the time series case and provided both theory and practical applications.

In this subsection, we briefly mention the selection procedure for bandwidth for the case where the weight function is a kernel weight.

For $1 \le i \le n$, define

$$\begin{aligned} \widetilde{\omega}_{i,n}(t) &= K\Big(\frac{t-T_i}{h}\Big)\Big/ \sum_{j=1,j\neq i}^{n} K\Big(\frac{t-T_j}{h}\Big), \\ \widetilde{g}_{i,n}(t,\beta) &= \sum_{j=1,j\neq i}^{n} \widetilde{\omega}_{j,n}(t)(Y_j - X_j^T\beta). \end{aligned}$$

We now define the modified LS estimator $\widetilde{\beta}(h)$ of β by minimizing

$$\sum_{i=1}^{n}\{Y_i - X_i^T\beta - \widetilde{g}_{i,n}(T_i,\beta)\}^2.$$

The Cross-Validation (CV) function can be defined as

$$CV(h) = \frac{1}{n}\sum_{i=1}^{n}\{Y_i - X_i^T\widetilde{\beta}(h) - \widetilde{g}_{i,n}(T_i,\widetilde{\beta}(h))\}^2.$$

Let $\widehat{h}$ denote the estimator of h, which is obtained by minimizing the CV function CV(h) over $h \in \Theta_h$, where Θ_h is an interval defined by

$$\Theta_h = [\lambda_1 n^{-1/5-\eta_1}, \lambda_2 n^{-1/5+\eta_1}],$$

where $0 < \lambda_1 < \lambda_2 < \infty$ and $0 < \eta_1 < 1/20$ are constants. Under Assumptions 1.3.1-1.3.3, we can show that the CV function provides an optimum bandwidth for estimating both β and g. Details for the i.i.d. case are similar to those in Section 6.4.

TABLE 2.1. Simulation results ($\times 10^{-3}$)

Estimator	Variance Model	$\beta_0 = 1$ Bias	$\beta_0 = 1$ MSE	$\beta_1 = 0.75$ Bias	$\beta_1 = 0.75$ MSE
LSE	1	8.696	8.7291	23.401	9.1567
WLSE	1	4.230	2.2592	1.93	2.0011
LSE	2	12.882	7.2312	5.595	8.4213
WLSE	2	5.676	1.9235	0.357	1.3241
LSE	3	5.9	4.351	18.83	8.521
WLSE	3	1.87	1.762	3.94	2.642

2.1.4 Simulation Comparisons

We present a small simulation study to illustrate the properties of the theoretical results in this chapter. We consider the following model with different variance functions.

$$Y_i = X_i^T \beta + g(T_i) + \sigma_i \varepsilon_i, \qquad i = 1, \ldots, n = 300,$$

where $\{\varepsilon_i\}$ is a sequence of the standard normal random variables, $\{X_i\}$ and $\{T_i\}$ are mutually independent uniform random variables on $[0,1]$, $\beta = (1, 0.75)^T$ and $g(t) = \sin(t)$. The number of simulations for each situation is 500.

Three models for the variance functions are considered. LSE and WLSE represent the least squares estimator and the **weighted least squares** estimator given in (1.1.1) and (2.1.5), respectively.

- Model 1: $\sigma_i^2 = T_i^2$;
- Model 2: $\sigma_i^2 = W_i^3$; where W_i are i.i.d. uniformly distributed random variables.
- Model 3: $\sigma_i^2 = a_1 \exp[a_2\{X_i^T\beta + g(T_i)\}^2]$, where $(a_1, a_2) = (1/4, 1/3200)$. The case where $g \equiv 0$ has been studied by Carroll (1982).

From Table 2.1, one can find that our estimator (WLSE) is better than LSE in the sense of both bias and MSE for each of the models.

By the way, we also study the behavior of the estimate for the nonparametric part $g(t)$

$$\sum_{i=1}^{n} \omega_{ni}^*(t)(\widetilde{Y}_i - \widetilde{X}_i^T \beta_{WLS}),$$

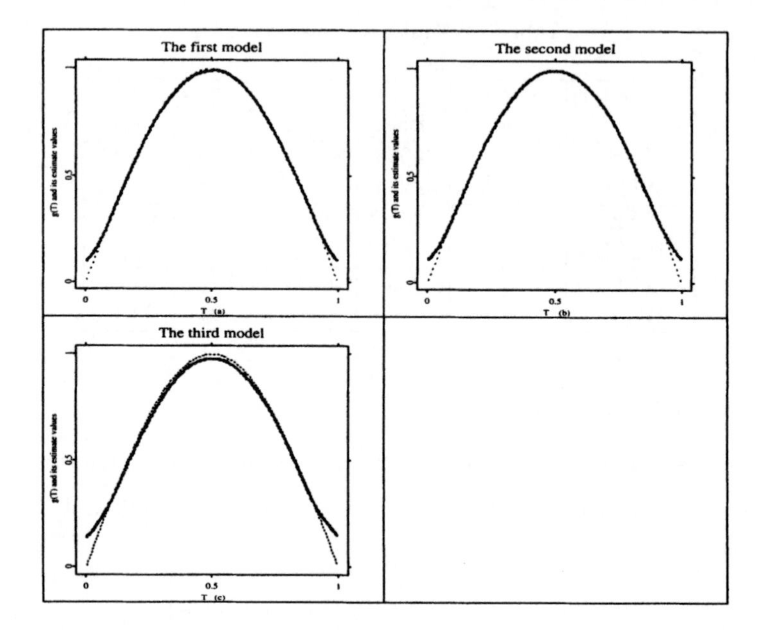

FIGURE 2.2. Estimates of the function $g(T)$ for the three models

where $\omega_{ni}^*(\cdot)$ are weight functions satisfying Assumption 1.3.3. In simulation, we take `Nadaraya-Watson weight` function with `quartic kernel`$(15/16)(1 - u^2)^2 I(|u| \leq 1)$ and use the `cross-validation` criterion to select the bandwidth. Figure 2.2 presents for the simulation results of the nonparametric parts of models 1–3, respectively. In the three pictures, thin dashed lines stand for true values and thick solid lines for our estimate values. The figures indicate that our estimators for the nonparametric part perform also well except in the neighborhoods of the points 0 and 1.

2.1.5 Technical Details

We introduce the following notation,

$$\widehat{A}_n = \sum_{i=1}^n \widehat{\gamma}_i \widetilde{X}_i \widetilde{X}_i^T, \quad A_n = \sum_{i=1}^n \gamma_i \widetilde{X}_i \widetilde{X}_i^T.$$

Proof of Theorem 2.1.2. It follows from the definition of β_W that

$$\beta_W - \beta = A_n^{-1} \Big\{ \sum_{i=1}^n \gamma_i \widetilde{X}_i \widetilde{g}(T_i) + \sum_{i=1}^n \gamma_i \widetilde{X}_i \widetilde{\varepsilon}_i \Big\}.$$

We will complete the proof by proving the following three facts for $j = 1, \ldots, p$,

(i) $H_{1j} = 1/\sqrt{n}\sum_{i=1}^n \gamma_i \widetilde{x}_{ij}\widetilde{g}(T_i) = o_P(1)$;

(ii) $H_{2j} = 1/\sqrt{n}\sum_{i=1}^n \gamma_i \widetilde{x}_{ij}\Big\{\sum_{k=1}^n \omega_{nk}(T_i)\xi_k\Big\} = o_P(1)$;

(iii) $H_3 = 1/\sqrt{n}\sum_{i=1}^n \gamma_i \widetilde{X}_i\xi_i \longrightarrow^{\mathcal{L}} N(0, B^{-1}\Sigma B^{-1})$.

The proof of (i) is mainly based on Lemmas A.1 and A.3. Observe that

$$\sqrt{n}H_{1j} = \sum_{i=1}^n \gamma_i u_{ij}\widetilde{g}_i + \sum_{i=1}^n \gamma_i h_{nij}\widetilde{g}_i - \sum_{i=1}^n \gamma_i \sum_{q=1}^n \omega_{nq}(T_i)u_{qj}\widetilde{g}_i, \tag{2.1.15}$$

where $h_{nij} = h_j(T_i) - \sum_{k=1}^n \omega_{nk}(T_i)h_j(T_k)$. In Lemma A.3, we take $r = 2$, $V_k = u_{kl}$, $a_{ji} = \widetilde{g}_j$, $1/4 < p_1 < 1/3$ and $p_2 = 1 - p_1$. Then, the first term of (2.1.15) is

$$O_P(n^{-(2p_1-1)/2}) = o_P(n^{1/2}).$$

The second term of (2.1.15) can be easily shown to be order $O_P(nc_n^2)$ by using Lemma A.1.

The proof of the third term of (2.1.15) follows from Lemmas A.1 and A.3, and

$$\begin{aligned}\Big|\sum_{i=1}^n\sum_{q=1}^n \gamma_i\omega_{nq}(T_i)u_{qj}\widetilde{g}_i\Big| &\le C_2 n \max_{i\le n}|\widetilde{g}_i| \max_{i\le n}\Big|\sum_{q=1}^n \omega_{nq}(T_i)u_{qj}\Big| \\ &= O(n^{2/3}c_n \log n) = o_p(n^{1/2}).\end{aligned}$$

Thus we complete the proof of (i).

We now show (ii), i.e., $\sqrt{n}H_{2j} \to 0$. Notice that

$$\begin{aligned}\sqrt{n}H_{2j} &= \sum_{i=1}^n \gamma_i\Big\{\sum_{k=1}^n \widetilde{x}_{kj}\omega_{ni}(T_k)\Big\}\xi_i \\ &= \sum_{i=1}^n \gamma_i\Big\{\sum_{k=1}^n u_{kj}\omega_{ni}(T_k)\Big\}\xi_i + \sum_{i=1}^n \gamma_i\Big\{\sum_{k=1}^n h_{nkj}\omega_{ni}(T_k)\Big\}\xi_i \\ &\quad -\sum_{i=1}^n \gamma_i\Big[\sum_{k=1}^n\Big\{\sum_{q=1}^n u_{qj}\omega_{nq}(T_k)\Big\}\omega_{ni}(T_k)\Big]\xi_i. \end{aligned}\tag{2.1.16}$$

The order of the first term of (2.1.16) is $O(n^{-(2p_1-1)/2}\log n)$ by letting $r = 2$, $V_k = \xi_k$, $a_{li} = \sum_{k=1}^n u_{kj}\omega_{ni}(T_k)$, $1/4 < p_1 < 1/3$ and $p_2 = 1 - p_1$ in Lemma A.3.

It follows from Lemma A.1 and (A.3) that the second term of (2.1.16) is bounded by

$$\begin{aligned}\Big|\sum_{i=1}^n \gamma_i\Big\{\sum_{k=1}^n h_{nkj}\omega_{ni}(T_k)\Big\}\xi_i\Big| &\le n\max_{k\le n}\Big|\sum_{i=1}^n \omega_{ni}(T_k)\xi_i\Big| \max_{j,k\le n}|h_{nkj}| \\ &= O(n^{2/3}c_n\log n) \quad a.s. \end{aligned}\tag{2.1.17}$$

The same argument as that for (2.1.17) yields that the third term of (2.1.16) is bounded by

$$\begin{aligned}
&\Big|\sum_{k=1}^{n}\Big\{\sum_{i=1}^{n}\gamma_i\omega_{ni}(T_k)\xi_i\Big\}\Big\{\sum_{q=1}^{n}u_{qj}\omega_{nq}(T_k)\Big\}\Big| \\
\leq\ & n\max_{k\leq n}\Big|\sum_{i=1}^{n}\omega_{ni}(T_k)\xi_i\Big| \times \max_{k\leq n}\Big|\sum_{q=1}^{k}u_{qj}\omega_{nq}(T_j)\Big| \\
=\ & O_P(n^{1/3}\log^2 n) = o_P(n^{1/2}). \qquad (2.1.18)
\end{aligned}$$

A combination (2.1.16)–(2.1.18) implies (ii).

Using the same procedure as in Lemma A.2, we deduce that

$$\lim_{n\to\infty}\frac{1}{n}\sum_{i=1}^{n}\gamma_i\widetilde{X}_i^T\widetilde{X}_i = B. \qquad (2.1.19)$$

A central limit theorem shows that as $n \to \infty$

$$\frac{1}{\sqrt{n}}\sum_{i=1}^{n}\gamma_i\widetilde{X}_i\xi_i \longrightarrow^{\mathcal{L}} N(0,\Sigma).$$

We therefore show that as $n \to \infty$

$$\frac{1}{\sqrt{n}}A_n^{-1}\sum_{i=1}^{n}\gamma_i\widetilde{X}_i\xi_i \longrightarrow^{\mathcal{L}} N(0,B^{-1}\Sigma B^{-1}).$$

This completes the proof of Theorem 2.1.2.

Proof of Theorem 2.1.3. In order to complete the proof of Theorem 2.1.3, we only need to prove

$$\sqrt{n}(\beta_{WLS} - \beta_W) = o_P(1).$$

First we state a fact, whose proof is immediately derived by (2.1.6) and (2.1.19),

$$\frac{1}{n}|\widehat{a}_n(j,l) - a_n(j,l)| = o_P(n^{-q}) \qquad (2.1.20)$$

for $j,l = 1,\ldots,p$, where $\widehat{a}_n(j,l)$ and $a_n(j,l)$ are the $(j,l)-$th elements of $\widehat{A}_n$ and A_n, respectively. The fact (2.1.20) will be often used later.

It follows that

$$\begin{aligned}
\beta_{WLS} - \beta_W =\ & \frac{1}{2}\Big\{A_n^{-1}(A_n - \widehat{A}_n)\widehat{A}_n^{-1}\sum_{i=1}^{n}\gamma_i\widetilde{X}_i\widetilde{g}(T_i) \\
& + \widehat{A}_n^{-1}\sum_{i=1}^{k_n}(\gamma_i - \widehat{\gamma}_i^{(2)})\widetilde{X}_i\widetilde{g}(T_i) + \widehat{A}_n^{-1}(A_n - \widehat{A}_n)\widehat{A}_n^{-1}\sum_{i=1}^{n}\gamma_i\widetilde{X}_i\widetilde{\xi}_i \\
& + \widehat{A}_n^{-1}\sum_{i=1}^{k_n}(\gamma_i - \widehat{\gamma}_i^{(2)})\widetilde{X}_i\widetilde{\xi}_i + \widehat{A}_n^{-1}\sum_{i=k_n+1}^{n}(\gamma_i - \widehat{\gamma}_i^{(1)})\widetilde{X}_i\widetilde{g}(T_i) \\
& + \widehat{A}_n^{-1}\sum_{i=k_n+1}^{n}(\gamma_i - \widehat{\gamma}_i^{(1)})\widetilde{X}_i\widetilde{\xi}_i\Big\}. \qquad (2.1.21)
\end{aligned}$$

By `Cauchy-Schwarz inequality`, for any $j = 1, \ldots, p$,

$$\Big|\sum_{i=1}^{n} \gamma_i \widetilde{x}_{ij} \widetilde{g}(T_i)\Big| \le C\sqrt{n} \max_{i \le n} |\widetilde{g}(T_i)| \Big(\sum_{i=1}^{n} \widetilde{x}_{ij}^2\Big)^{1/2},$$

which is $o_P(n^{3/4})$ by Lemma A.1 and (2.1.19). Thus each element of the first term of (2.1.21) is $o_P(n^{-1/2})$ by using the fact that each element of $A_n^{-1}(A_n - \widehat{A}_n)\widehat{A}_n^{-1}$ is $o_P(n^{-5/4})$. The similar argument demonstrates that each element of the second and fifth terms is also $o_P(n^{-1/2})$.

Similar to the proof of $H_{2j} = o_P(1)$, and using the fact that H_3 converges to the normal distribution, we conclude that the third term of (2.1.21) is also $o_P(n^{-1/2})$. It suffices to show that the fourth and the last terms of (2.1.21) are both $o_P(n^{-1/2})$. Since their proofs are the same, we only show that for $j = 1, \ldots, p$,

$$\Big\{\widehat{A}_n^{-1} \sum_{i=1}^{k_n} (\gamma_i - \widehat{\gamma}_i^{(2)}) \widetilde{X}_i \widetilde{\xi}_i\Big\}_j = o_P(n^{-1/2})$$

or equivalently

$$\sum_{i=1}^{k_n} (\gamma_i - \widehat{\gamma}_i^{(2)}) \widetilde{x}_{ij} \widetilde{\xi}_i = o_P(n^{1/2}). \tag{2.1.22}$$

Let $\{\delta_n\}$ be a sequence of real numbers converging to zero but satisfying $\delta_n > n^{-1/4}$. Then for any $\mu > 0$ and $j = 1, \ldots, p$,

$$\begin{aligned} P\Big\{\Big|\sum_{i=1}^{k_n} (\gamma_i &- \widehat{\gamma}_i^{(2)}) \widetilde{x}_{ij} \xi_i I(|\gamma_i - \widehat{\gamma}_i^{(2)}| \ge \delta_n)\Big| > \mu n^{1/2}\Big\} \\ &\le P\Big\{\max_{i \le n} |\gamma_i - \widehat{\gamma}_i^{(2)}| \ge \delta_n\Big\} \to 0. \end{aligned} \tag{2.1.23}$$

The last step is due to (2.1.6).

Next we deal with the term

$$P\Big\{\Big|\sum_{i=1}^{k_n} (\gamma_i - \widehat{\gamma}_i^{(2)}) \widetilde{x}_{ij} \xi_i I(|\gamma_i - \widehat{\gamma}_i^{(2)}| \le \delta_n)\Big| > \mu n^{1/2}\Big\}$$

using `Chebyshev's inequality`. Since $\widehat{\gamma}_i^{(2)}$ are independent of ξ_i for $i = 1, \ldots, k_n$, we can easily derive

$$E\Big\{\sum_{i=1}^{k_n} (\gamma_i - \widehat{\gamma}_i^{(2)}) \widetilde{x}_{ij} \xi_i\Big\}^2 = \sum_{i=1}^{k_n} E\{(\gamma_i - \widehat{\gamma}_i^{(2)}) \widetilde{x}_{ij} \xi_i\}^2.$$

This is why we use the `split-sample` technique to estimate γ_i by $\widehat{\gamma}_i^{(2)}$ and $\widehat{\gamma}_i^{(1)}$. In fact,

$$P\Big\{\Big|\sum_{i=1}^{k_n} (\gamma_i - \widehat{\gamma}_i^{(2)}) \widetilde{x}_{ij} \xi_i I(|\gamma_i - \widehat{\gamma}_i^{(2)}| \le \delta_n)\Big| > \mu n^{1/2}\Big\}$$

$$\leq \frac{\sum_{i=1}^{k_n} E\{(\gamma_i - \widehat{\gamma}_i^{(2)})I(|\gamma_i - \widehat{\gamma}_i^{(2)}| \leq \delta_n)\}^2 E\|\widetilde{X}_i\|^2 E\xi_i^2}{n\mu^2}$$
$$\leq C\frac{k_n\delta_n^2}{n\mu^2} \to 0. \qquad (2.1.24)$$

Thus, by (2.1.23) and (2.1.24),

$$\sum_{i=1}^{k_n}(\gamma_i - \widehat{\gamma}_i^{(2)})\widetilde{x}_{ij}\xi_i = o_P(n^{1/2}).$$

Finally,

$$\Big|\sum_{i=1}^{k_n}(\gamma_i - \widehat{\gamma}_i^{(2)})\widetilde{x}_{ij}\Big\{\sum_{k=1}^{n}\omega_{nk}(T_i)\xi_k\Big\}\Big|$$
$$\leq \sqrt{n}\Big(\sum_{i=1}^{k_n}\widetilde{X}_{ij}^2\Big)^{1/2}\max_{1\leq i\leq n}|\gamma_i - \widehat{\gamma}_i^{(2)}|\max_{1\leq i\leq n}\Big|\sum_{k=1}^{n}\omega_{nk}(T_i)\xi_k\Big|.$$

This is $o_P(n^{1/2})$ by using (2.1.20), (A.3), and (2.1.19). Therefore, we complete the proof of Theorem 2.1.3.

2.2 Estimation with Censored Data

2.2.1 Introduction

We are here interested in the estimation of β in model (1.1.1) when the response Y_i are incompletely observed and right-**censored** by random variables Z_i. That is, we observe

$$Q_i = \min(Z_i, Y_i), \quad \delta_i = I(Y_i \leq Z_i), \qquad (2.2.1)$$

where Z_i are i.i.d., and Z_i and Y_i are mutually independent. We assume that Y_i and Z_i have a common distribution F and an unknown distribution G, respectively. In this section, we also assume that ε_i are i.i.d. and that (X_i, T_i) are random designs and that (Z_i, X_i^T) are independent random vectors and independent of the sequence $\{\varepsilon_i\}$. The main results are based on the paper of Liang and Zhou (1998).

When the Y_i are observable, the estimator of β with the ordinary rate of convergence is given in (1.2.2). In present situation, the least squares form of (1.2.2) cannot be used any more since Y_i are not observed completely. It is well-known that in linear and nonlinear **censored** regression models, consistent estimators are obtained by replacing incomplete observations with **synthetic data**. See, for example, Buckley and James (1979), Koul, Susarla and Ryzin (1981), Lai and Ying (1991, 1992) and Zhou (1992). In our context, these suggest that we use the following estimator

$$\widehat{\beta}_n = \Big[\sum_{i=1}^{n}\{X_i - \widehat{g}_{x,h}(T_i)\}^{\otimes 2}\Big]^{-1} \sum_{i=1}^{n}\{X_i - \widehat{g}_{x,h}(T_i)\}\{Y_i^* - \widehat{g}_{y^*,h}(T_i)\} \qquad (2.2.2)$$

for some `synthetic data` Y_i^*, where $A^{\otimes 2} \stackrel{\text{def}}{=} A \times A^T$.

In this section, we analyze the estimate (2.2.2) and show that it is asymptotically normal for appropriate `synthetic data` Y_i^*.

2.2.2 Synthetic Data and Statement of the Main Results

We assume that G is known first. The unknown case is discussed in the second part. The third part states the main results.

2.2.2.1 When G is Known

Define `synthetic data`

$$Y_{i(1)} = \phi_1(Q_i, G)\delta_i + \phi_2(Q_i, G)(1 - \delta_i), \qquad (2.2.3)$$

where ϕ_1 and ϕ_2 are continuous functions which satisfy

(i). $\{1 - G(Y)\}\phi_1(Y, G) + \int_{-\infty}^{Y} \phi_2(t, G)dG(t) = Y$;

(ii). ϕ_1 and ϕ_2 don't depend on F.

The set containing all pairs (ϕ_1, ϕ_2) satisfying (i) and (ii) is denoted by $\mathcal{K}$.

Remark 2.2.1 *Equation (2.2.3) plays an important role in our case. Note that* $E(Y_{i(1)}|T_i, X_i) = E(Y_i|T_i, X_i)$ *by (i), which implies that the regressors of* $Y_{i(1)}$ *and* Y_i *on* (W, X) *are the same. In addition, if* $Z = \infty$, *or* Y_i *are completely observed, then* $Y_{i(1)} = Y_i$ *by taking* $\phi_1(u, G) = u/\{1 - G(u)\}$ *and* $\phi_2 = 0$. *So our* `synthetic data` *are the same as the original ones.*

Remark 2.2.2 *We here list the variances of the* `synthetic data` *for the following three pairs* (ϕ_1, ϕ_2). *Their calculations are direct and we therefore omit the details.*

- $\phi_1(u,G) = u$, $\phi_2(u,G) = u + G(u)/G'(u)$,

$$Var(Y_{i(1)}) = Var(Y) + \int_0^\infty \left\{ \frac{G(u)}{G'(u)} \right\}^2 \{1 - F(u)\} dG(u).$$

- $\phi_1(u,G) = u/\{1 - G(u)\}$, $\phi_2 = 0$,

$$Var(Y_{i(1)}) = Var(Y) + \int_0^\infty \frac{u^2 G(u)}{1 - G(u)} dF(u).$$

- $\phi_1(u,G) = \phi_2(u,G) = \int_{-\infty}^u \{1 - G(s)\}^{-1} ds$,

$$Var(Y_{i(1)}) = Var(Y) + 2\int_0^\infty \{1 - F(u)\} \int_0^u \frac{G(s)}{1 - G(s)} ds dG(u).$$

These arguments indicate that each of the variances of $Y_{i(1)}$ is greater than that of Y_i, which is pretty reasonable since we have modified Y_i. We cannot compare the variances for different (ϕ_1, ϕ_2), which depends on the behavior of $G(u)$. Therefore, it is difficult to recommend the choice of (ϕ_1, ϕ_2) absolutely.

Equation (2.2.2) suggests that the generalized least squares estimator of β is

$$\beta_{n(1)} = (\widetilde{\mathbf{X}}^T \widetilde{\mathbf{X}})^{-1} (\widetilde{\mathbf{X}}^T \widetilde{\mathbf{Y}}_{(1)}) \tag{2.2.4}$$

where $\widetilde{\mathbf{Y}}_{(1)}$ denotes $(\widetilde{Y}_{1(1)}, \ldots, \widetilde{Y}_{n(1)})$ with $\widetilde{Y}_{i(1)} = Y_{i(1)} - \sum_{j=1}^n \omega_{nj}(T_i) Y_{j(1)}$.

2.2.2.2 When G is Unknown

Generally, $G(\cdot)$ is unknown in practice and must be estimated. The usual estimate of $G(\cdot)$ is a modification of its `Kaplan-Meier estimator`. In order to construct our estimators, we need to assume $G(\sup_{(X,W)} T_{F(X,W)}) \le \gamma$ for some known $0 < \gamma < 1$, where $T_{F(X,W)} = \inf\{y; F_{(X,W)}(y) = 1\}$ and $F_{(X,W)}(y) = P\{Y \le y | X, W\}$. Let $1/3 < \nu < 1/2$ and $\tau_n = \sup\{t : 1 - F(t) \ge n^{-(1-\nu)}\}$. Then a simple modification of the `Kaplan-Meier estimator` is

$$G_n^\Delta(z) = \begin{cases} \widehat{G}_n(z), & \text{if } \widehat{G}_n(z) \le \gamma, \\ \\ \gamma, & \text{if } z \le \max Q_i \text{ and } \widehat{G}_n(z) > \gamma, \end{cases} \qquad i = 1, \ldots, n$$

where $\widehat{G}_n(z)$ is the `Kaplan-Meier estimator` given by

$$\widehat{G}_n(z) = 1 - \prod_{Q_i \le z} \Big(1 - \frac{1}{n - i + 1}\Big)^{(1-\delta_i)}.$$

Substituting G in (2.2.3) by G_n^Δ, we get the **synthetic data** for the case of unknown $G(u)$, that is,

$$Y_{i(2)} = \phi_1(Q_i, G_n^\Delta)\delta_i + \phi_2(Q_i, G_n^\Delta)(1-\delta_i).$$

Replacing $Y_{i(1)}$ in (2.2.4) by $Y_{i(2)}$, we get an estimate of β for the case of unknown $G(\cdot)$. For convenience, we make a modification of (2.2.4) by employing the **split-sample** technique as follows. Let $k_n (\le n/2)$ be the largest integer part of $n/2$. Let $G_{n1}^\Delta(\bullet)$ and $G_{n2}^\Delta(\bullet)$ be the estimators of G based on the observations $(Q_1, \ldots, Q_{k_n})$ and $(Q_{k_n+1}, \ldots, Q_n)$, respectively. Denote

$$Y_{i(2)}^{(1)} = \phi_1(Q_i, G_{n2}^\Delta)\delta_i + \phi_2(Q_i, G_{n2}^\Delta)(1-\delta_i) \text{ for } i = 1, \ldots, k_n$$

and

$$Y_{i(2)}^{(2)} = \phi_1(Q_i, G_{n1}^\Delta)\delta_i + \phi_2(Q_i, G_{n1}^\Delta)(1-\delta_i) \text{ for } i = k_n+1, \ldots, n.$$

Finally, we define

$$\beta_{n(2)} = (\widetilde{\mathbf{X}}^T\widetilde{\mathbf{X}})^{-1}\Big\{\sum_{i=1}^{k_n} \widetilde{X}_i^{(1)}\widetilde{Y}_{i(2)}^{(1)} + \sum_{i=k_n+1}^{n} \widetilde{X}_i^{(2)}\widetilde{Y}_{i(2)}^{(2)}\Big\}$$

as the estimator of β and modify the estimator given in (2.2.4) as

$$\beta_{n(1)} = (\widetilde{\mathbf{X}}^T\widetilde{\mathbf{X}})^{-1}\Big\{\sum_{i=1}^{k_n} \widetilde{X}_i^{(1)}\widetilde{Y}_{i(1)}^{(1)} + \sum_{i=k_n+1}^{n} \widetilde{X}_i^{(2)}\widetilde{Y}_{i(1)}^{(2)}\Big\},$$

where

$$\widetilde{Y}_{i(2)}^{(1)} = Y_{i(2)}^{(1)} - \sum_{j=1}^{k_n}\omega_{nj}(T_i)Y_{j(2)}^{(1)}, \quad \widetilde{Y}_{i(2)}^{(2)} = Y_{i(2)}^{(2)} - \sum_{j=k_n+1}^{n}\omega_{nj}(T_i)Y_{j(2)}^{(2)}$$

$$\widetilde{Y}_{i(1)}^{(1)} = Y_{i(1)} - \sum_{j=1}^{k_n}\omega_{nj}(T_i)Y_{j(1)}, \quad \widetilde{Y}_{i(1)}^{(2)} = Y_{i(1)} - \sum_{j=k_n+1}^{n}\omega_{nj}(T_i)Y_{j(1)}$$

$$\widetilde{X}_i^{(1)} = X_i - \sum_{j=1}^{k_n}\omega_{nj}(T_i)X_j, \quad \widetilde{X}_i^{(2)} = X_i - \sum_{j=k_n+1}^{n}\omega_{nj}(T_i)X_j.$$

2.2.2.3 Main Results

Theorem 2.2.1 *Suppose that Assumptions 1.3.1-1.3.3 hold. Let* $(\phi_1, \phi_2) \in \mathcal{K}$ *and* $E|X|^4 < \infty$. *Then* $\beta_{n(1)}$ *is an asymptotically normal estimator of* β, *that is,*

$$n^{1/2}(\beta_{n(1)} - \beta) \longrightarrow^{\mathcal{L}} N(0, \Sigma^*)$$

where $\Sigma^* = \Sigma^{-2}E\{\epsilon_{1(1)}^2 u_1 u_1^T\}$ *with* $\epsilon_{1(1)} = Y_{1(1)} - E\{Y_{1(1)}|X, W\}$.

Let $\mathcal{K}^*$ be a subset of $\mathcal{K}$, consisting of all the elements (ϕ_1, ϕ_2) satisfying the following: There exists a constant $0 < \mathcal{C} < \infty$ such that

$$\max_{j=1,2,u\leq s} |\phi_j(u,G)| < \mathcal{C} \quad \text{for all s with } G(s) < 1$$

and there exist constants $0 < L = L(s) < \infty$ and $\eta > 0$ such that

$$\max_{j=1,2,u\leq s} |\phi_j(u,G^*) - \phi_j(u,G)| \leq L \sup_{u\leq s} |G^*(u) - G(u)|$$

for all distribution functions G^* with $\sup_{u\leq s} |G^*(u) - G(u)| \leq \eta$.

Assumption 2.2.1 *Assume that $F(w)$ and $G(w)$ are continuous. Let*

$$\int_{-\infty}^{T_F} \frac{1}{1-F(s)} dG(s) < \infty.$$

Theorem 2.2.2 *Under the conditions of Theorem 2.2.1 and Assumption 2.2.1, $\beta_{n(1)}$ and $\beta_{n(2)}$ have the same normal limit distribution with mean β and covariance matrix Σ^*.*

2.2.3 Estimation of the Asymptotic Variance

Subsection 2.2.2 gives asymptotic normal approximations to the estimators $\beta_{n(1)}$ and $\beta_{n(2)}$ with asymptotic variance Σ^*. In principle, Σ^* is unknown and must be estimated. The usual method is to replace $X - E(X|T)$ by $X_i - \Gamma_n(X_i)$ and define

$$\begin{aligned}
\Sigma_n &= \frac{1}{n}\sum_{i=1}^{n}\{X_i - \Gamma_n(X_i)\}^{\otimes 2}, \\
V_{n(2)} &= \frac{1}{n}\sum_{i=1}^{n}\Big[\{X_i - \Gamma_n(X_i)\}^{\otimes 2}\{Y_{i(2)} - X_i^T\beta_{n(2)} - g_n(T_i)\}^2\Big]
\end{aligned}$$

as the estimators of Σ and $E\{\epsilon_{1(1)}^2 u_1 u_1^T\}$, respectively, where

$$\begin{aligned}
\Gamma_n(X_i) &= \sum_{j=1}^{n} \omega_{nj}(T_i)X_j, \\
g_n(T_i) &= \sum_{j=1}^{n} \omega_{nj}(T_i)Y_{j(2)} - \sum_{j=1}^{n}\omega_{nj}(T_i)X_j^T\beta_{n(2)}.
\end{aligned}$$

Using the same techniques as in the proof of Theorem 2.2.2, one can show that Σ_n and $V_{n(2)}$ are consistent estimators of Σ and $E(\epsilon_{1(1)}^2 u_1 u_1^T)$, respectively. Hence, $\Sigma_n^{-2}E_{n(2)}$ is a consistent estimator of $\Sigma^{-2}E\{\epsilon_{1(1)}^2 u_1 u_1^T\}$.

2.2.4 A Numerical Example

To illustrate the behavior of our estimator $\beta_{n(2)}$, we present some small sample simulations to study the sample bias and mean squares error (MSE) of the estimate $\beta_{n(2)}$. We consider the model given by

$$Y_i = X_i^T \beta_0 + T_i^3 + \varepsilon_i, \qquad i = 1, \ldots, n \text{ for } n = 30, 50,$$

where X_i are i.i.d. with two-dimensional uniform distribution $U[0, 100; 0, 100]$, T_i are i.i.d. drawn from $U[0, 1]$, and $\beta_0 = (2, 1.75)^T$. The right-censored random variables Z_i are i.i.d. with $\exp(-0.008z)$, the exponential distribution function with freedom degree $\lambda = 0.008$, and ε_i are i.i.d. with common $N(0, 1)$ distribution. Three pairs of (ϕ_1, ϕ_2) are considered:

- P1: $\phi_1(u, G) = u$, $\phi_2(u, G) = u + G(u)/G'(u)$;
- P2: $\phi_1(u, G) = u/\{1 - G(u)\}$, $\phi_2 = 0$;
- P3: $\phi_1(u, G) = \phi_2(u, G) = \int_{-\infty}^{u} \{1 - G(u)\}^{-1} ds$.

The results in Table 2.2 are based on 500 replications. The simulation study shows that our estimation procedure works very well numerically in small sample case. It is worthwhile to mention that, after a direct calculation using Remark 2.2.2, the variance of the estimator based on the first pair (ϕ_1, ϕ_2) is the smallest one in our context, while that based on the second pair (ϕ_1, ϕ_2) is the largest one, with which the simulation results also coincide.

TABLE 2.2. Simulation results

n	MODELS	$\beta_1 = 2$	$\beta_1 = 1.75$
		MSE	MSE
30	P1	0.0166717	0.0170406
	P2	0.0569963	0.0541931
	P3	0.0191386	0.0180647
50	P1	0.0103607	0.0099157
	P2	0.0277258	0.0268500
	P3	0.0129026	0.0118281

2.2.5 Technical Details

Since $T_{F_{(X,W)}} < T_G < \infty$ for $T_G = \inf\{z; G(z) = 1\}$ and $P(Q > \tau_n|\mathbf{X}, \mathbf{W}) = n^{-(1-\nu)}$, we have $P(Q > \tau_n) = n^{-(1-\nu)}$, $1 - F(\tau_n) \geq n^{-(1-\nu)}$ and $1 - G(\tau_n) \geq n^{-(1-\nu)}$. These will be often used later.

The proof of Theorem 2.2.1. The proof of the fact that $n^{1/2}(\beta_{n(1)} - \beta)$ converges to $N(0, \Sigma^*)$ in distribution can be completed by slightly modifying the proof of Theorem 3.1 of Zhou (1992). We therefore omit the details.

Before proving Theorem 2.2.2, we state the following two lemmas.

Lemma 2.2.1 *If $E|X|^4 < \infty$ and $(\phi_1, \phi_2) \in \mathcal{K}$, then for some given positive integer $m \leq 4$, $E(|Y_{i(1)}|^m \Big| \mathbf{X}, \mathbf{W}) \leq C$, $E\{\epsilon_{1(1)}^2 u_1 u_1^T\}^2 < \infty$ and $E(|Y_{i(1)}|^m I_{(Q_i > \tau)} \Big| \mathbf{X}, \mathbf{W}) \leq CP(Q_i > \tau \Big| \mathbf{X}, \mathbf{W})$ for any $\tau \in R^1$.*

Lemma 2.2.2 *(See Gu and Lai, 1990) Assume that Assumption 2.2.1 holds. Then*

$$\limsup_{n\to\infty} \sqrt{\frac{n}{2\log_2 n}} \sup_{z \leq \tau_n} |\widehat{G}_n(z) - G(z)| = \sup_{z \leq T_F} \sqrt{\frac{S(z)}{\sigma(z)}}, \quad a.s.,$$

where $S(z) = 1 - G(z)$, $\sigma(z) = \int_{-\infty}^{z} S^{-2}(s)\{1 - F(s)\}^{-1} dG(s)$, and $\log_2$ denotes $\log\log$.

The proof of Theorem 2.2.2. We shall show that $\sqrt{n}(\beta_{n(2)} - \beta_{n(1)})$ converges to zero in probability. For the sake of simplicity of notation, we suppose $k = 1$ without loss of generality and denote $h(t) = E(X_1|T_1 = t)$ and $u_i = X_i - h(T_i)$ for $i = 1, \ldots, n$.

Denote $n(\widetilde{\mathbf{X}}^T \widetilde{\mathbf{X}})^{-1}$ by $A(n)$. By a direct calculation, $\sqrt{n}(\beta_{n(2)} - \beta_{n(1)})$ can be decomposed as follows:

$$A(n) n^{-1/2} \sum_{i=1}^{k_n} \widetilde{X}_i^{(1)} (\widetilde{Y}_{i(2)}^{(1)} - \widetilde{Y}_{i(1)}^{(1)}) + A(n) n^{-1/2} \sum_{i=k_n+1}^{n} \widetilde{X}_i^{(2)} (\widetilde{Y}_{i(2)}^{(2)} - \widetilde{Y}_{i(1)}^{(2)}).$$

It suffices to show that each of the above terms converges to zero in probability. Since their proofs are the same, we only show this assertion for the first term, which can be decomposed into

$$A(n) n^{-1/2} \sum_{i=1}^{k_n} \widetilde{h}(T_i) (\widetilde{Y}_{i(2)}^{(1)} - \widetilde{Y}_{i(1)}^{(1)}) I_{(Q_i \leq \tau_n)}$$
$$+ A(n) n^{-1/2} \sum_{i=1}^{k_n} \widetilde{u}_i (\widetilde{Y}_{i(2)}^{(1)} - \widetilde{Y}_{i(1)}^{(1)}) I_{(Q_i \leq \tau_n)}$$

$$+A(n)n^{-1/2}\sum_{i=1}^{k_n}\widetilde{X}_i(\widetilde{Y}_{i(2)}^{(1)}-\widetilde{Y}_{i(1)}^{(1)})I_{(Q_i>\tau_n)}$$
$$\stackrel{\text{def}}{=} A(n)n^{-1/2}(J_{n1}+J_{n2}+J_{n3}).$$

Lemma A.2 implies that $A(n)$ converges to Σ^{-1} in probability. Hence we only need to prove that these three terms are of $o_P(n^{1/2})$.

Obviously, Lemma 2.2.2 implies that

$$\sup_{t\le\tau_n}|\widehat{G}_n(t)-G(t)|=O(n^{-1/2}\log_2 n), \quad a.s. \tag{2.2.5}$$

Lemma A.1 implies that $\sup_i|\widetilde{h}(T_i)| = O(c_n)$. These arguments and a simple calculation demonstrate that

$$|J_{n1}|\le C\sup_i|\widetilde{h}(T_i)|\sum_{i=1}^{k_n}\Big\{1+\sum_{j=1}^{k_n}\omega_{nj}(T_i)\Big\}\sup_{t\le\tau_n}|\widehat{G}_n(t)-G(t)|$$
$$\le Cc_n n^{1/2}(\log_2 n)^{1/2}=o(n^{1/2}).$$

Analogously, J_{n2} is bounded by

$$\Big|\sum_{i=1}^{k_n}u_i(Y_{i(2)}^{(1)} - Y_{i(1)}^{(1)})I_{(Q_i\le\tau_n)}\Big|+\sum_{i=1}^{k_n}\Big|\sum_{j=1}^{k_n}\omega_{nj}(T_i)u_j(Y_{i(2)}^{(1)}-Y_{i(1)}^{(1)})\Big|I_{(Q_i\le\tau_n)}$$
$$= J_{n21}+J_{n22}. \tag{2.2.6}$$

Taking $V_i=u_i$ and $a_{nj}=\omega_{nj}(T_i)$ in Lemma A.3, we obtain

$$\sum_{k=1}^{k_n}\omega_{nk}(T_i)u_k=O(n^{-1/4}\log^{1/4}n), \tag{2.2.7}$$

which implies that

$$J_{n22} \le Cn^{-1/4}\log^{1/4}n\sum_{i=1}^{n}|Y_{(i(2)}^{(1)}-Y_{i(1)}^{(1)}|$$
$$\le Cn^{1/4}\log^{1/4}n=o(n^{1/2}) \quad a.s.$$

by the definitions of $Y_{i(2)}^{(1)}$ and $Y_{i(1)}^{(1)}$ and (2.2.5). We now show that

$$J_{n21}=o_P(n^{1/2}). \tag{2.2.8}$$

Since $Y_{i(2)}^{(1)}$ depend only on the first k_n-th samples, J_{n21} is the sum of the independent random variables given the last $n-k_n$ samples. From **Chebyshev's inequality**, for any given $\zeta>0$,

$$P\Big\{\Big|\sum_{i=1}^{k_n}u_i\Big\{Y_{i(2)}^{(1)}-Y_{i(1)}^{(1)}\Big\}I_{(Q_i\le\tau_n)}\Big|>\zeta n^{1/2}\Big\} \le \frac{1}{n\zeta^2}\sum_{i=1}^{k_n}Eu_i^2E(Y_{i(2)}^{(1)}-Y_{i(1)}^{(1)})^2$$
$$\le \frac{k_n}{n\zeta^2}Eu_1^2\sup_t|\widehat{G}_n(t)-G(t)|^2,$$

which converges to zero as n tends to infinite. Thus J_{n21} is $o_P(n^{1/2})$. A combination of the above arguments yields that $n^{-1/2}J_{n2}$ converges to zero in probability.

Next we show that $n^{-1/2}J_{n3}$ converges to zero in probability, which is equivalent to showing that the following sum converges to zero in probability,

$$n^{-1/2}\sum_{i=1}^{k_n}\widetilde{h}(T_i)(Y_{i(2)}^{(1)}-Y_{i(1)}^{(1)})I_{(Q_i>\tau_n)}+n^{-1/2}\sum_{i=1}^{k_n}\widetilde{u}_i(Y_{i(2)}^{(1)}-Y_{i(1)}^{(1)})I_{(Q_i>\tau_n)}. \quad (2.2.9)$$

The first term of (2.2.9) is bounded by

$$n^{-1/2}c_n\Big\{\max_i|Y_{i(2)}^{(1)}|\cdot\sum_{i=1}^{k_n}I_{(Q_i>\tau_n)}+\sum_{i=1}^{k_n}|Y_{n(1)}|I_{(Q_i>\tau_n)}\Big\},$$

which is bounded by $Cn^{-1/2}c_n\{nP(Q_1>\tau_n)+nE(|Y_{1(1)}|I_{(Q_1>\tau_n)})\}$ and then $o_P(1)$ by Lemma 2.2.1.

The second term of (2.2.9) equals

$$\begin{aligned}n^{-1/2}\sum_{i=1}^{k_n}u_i(Y_{i(2)}^{(1)} \quad &- \quad Y_{i(1)}^{(1)})I_{(Q_i>\tau_n)}\\ &- \quad n^{-1/2}\sum_{j=1}^{k_n}\Big\{\sum_{i=1}^{k_n}\omega_{nj}(T_i)(Y_{i(2)}^{(1)}-Y_{i(1)}^{(1)})I_{(Q_i>\tau_n)}\Big\}u_j. \quad (2.2.10)\end{aligned}$$

By Lemma 2.2.1, the first term of (2.2.10) is smaller than

$$n^{-1/2}\sum_{j=1}^{k_n}|u_jY_{i(2)}^{(1)}|I_{(Q_j\geq\tau_n)}+n^{-1/2}\sum_{j=1}^{k_n}|u_jY_{n(1)}|I_{(Q_j>\tau_n)},$$

which is further bounded by

$$\begin{aligned}n^{-1/2}\sum_{j=1}^{k_n}|u_j|I_{(Q_j\geq\tau_n)} \quad &+ \quad Cn^{1/2}E\{|u_1Y_{1(1)}|I_{(Q_1>\tau_n)}\}\\ &\leq \quad Cn^{1/2}E|u_1|I_{(Q_1>\tau_n)}\\ &= \quad Cn^{1/2}E\{|u_1|P(Q_1>\tau_n|X_1,T_1)\}\\ &= \quad O(n^{-1/2+\nu})=o_P(1). \quad (2.2.11)\end{aligned}$$

Similar to the proof of (2.2.11) and Lemma A.3, we can show that the second term of (2.2.10) is $o_P(1)$. We therefore complete the proof of Theorem 2.2.2.

2.3 Bootstrap Approximations

2.3.1 Introduction

The technique of bootstrap is a useful tool for the approximation to an unknown probability distribution as well as its characteristics like moments and confidence regions. In this section, we use the empirical distribution function to approximate the underlying error distribution (for more details see subsection 2.3.2). This classical bootstrap technique was introduced by Efron (for a review see e.g. Efron and Tibshirani, 1993 and Davison and Hinkley, 1997). Note that for a **heteroscedastic** error structure, a wild bootstrap procedure (see e.g. Wu (1986) or Härdle and Mammen (1993)) would be more appropriate.

Hong and Cheng (1993) considered using **bootstrap approximations** to the estimators of the parameters in model (1.1.1) for the case where $\{X_i, T_i, i = 1, \ldots, n\}$ is a sequence of i.i.d. random variables and $g(\cdot)$ is estimated by a kernel smoother. The authors proved that their **bootstrap approximations** are the same as the classic methods, but failed to explain the advantage of the bootstrap method. We will construct bootstrap statistics of β and σ^2, study their asymptotic normality when ε_i are i.i.d. and (X_i, T_i) are known design points, then show that the bootstrap techniques provide a reliable method to approximate the distributions of the estimates, and finally illustrate the method by a simulated example.

The effect of a smoothing parameter is discussed in a simulation study. Our research shows that the estimators of the parametric part are quite robust against the choice of the smoothing parameter. More details can be found in Subsection 2.3.3.

2.3.2 Bootstrap Approximations

In the partially linear model, the observable column $n-$vector $\hat{\varepsilon}$ of residuals is given by

$$\hat{\varepsilon} = \mathbf{Y} - \mathbf{G_n} - \mathbf{X}\beta_{LS},$$

where $\mathbf{G_n} = \{\widehat{g}_n(T_1), \ldots, \widehat{g}_n(T_n)\}^T$. Denote $\mu_n = 1/n \sum_{i=1}^n \hat{\varepsilon}_i$. Let $\hat{F}_n$ be the empirical distribution of $\hat{\varepsilon}$, centered at the mean, so $\hat{F}_n$ puts mass $1/n$ at $\hat{\varepsilon}_i - \mu_n$ and $\int x d\hat{F}_n(x) = 0$. Given $\mathbf{Y}$, let $\varepsilon_1^*, \ldots, \varepsilon_n^*$ be conditionally independent with the common distribution $\hat{F}_n$, ε^* be the $n-$vector whose $i-$th component is ε_i^*, and

$$\mathbf{Y}^* = \mathbf{X}\beta_{LS} + \mathbf{G_n} + \varepsilon^*,$$

where $\mathbf{Y}^*$ is generated from the data, β_{LS} is regarded as a vector of parameters. We denote the distribution of the disturbance terms ε^* by $\hat{F}_n$.

We now define the estimates of β and σ^2 by, respectively,

$$\beta_{LS}^* = (\widetilde{\mathbf{X}}^T\widetilde{\mathbf{X}})^{-1}\widetilde{\mathbf{X}}^T\widetilde{\mathbf{Y}}^* \quad \text{and} \quad \widehat{\sigma}_n^{2*} = \frac{1}{n}\sum_{i=1}^{n}(\widetilde{Y}_i^* - \widetilde{X}_i^T\beta_{LS}^*)^2,$$

where $\widetilde{\mathbf{Y}}^* = (\widetilde{Y}_1^*, \ldots, \widetilde{Y}_n^*)^T$ with $\widetilde{Y}_i^* = Y_i^* - \sum_{j=1}^n \omega_{nj}(T_i)Y_j^*$ for $i = 1, \ldots, n$.

The bootstrap principle asserts that the distributions of $\sqrt{n}(\beta_{LS}^* - \beta_{LS})$ and $\sqrt{n}(\widehat{\sigma}_n^{2*} - \widehat{\sigma}_n^2)$, which can be computed directly from the data, can approximate the distributions of $\sqrt{n}(\beta_{LS} - \beta)$ and $\sqrt{n}(\widehat{\sigma}_n^2 - \sigma^2)$, respectively. As shown later, this approximation works very well as $n \to \infty$. The main result of this section is given in the following theorem.

Theorem 2.3.1 *Suppose that Assumptions 1.3.1-1.3.3 hold. If* $\max_{1\leq i\leq n}\|u_i\| \leq C_0 < \infty$ *and* $E\varepsilon_1^4 < \infty$. *Then*

$$sup_x\left|P^*\{\sqrt{n}(\beta_{LS}^* - \beta_{LS}) < x\} - P\{\sqrt{n}(\beta_{LS} - \beta) < x\}\right| \to 0 \tag{2.3.1}$$

and

$$sup_x\left|P^*\{\sqrt{n}(\widehat{\sigma}_n^{2*} - \widehat{\sigma}_n^2) < x\} - P\{\sqrt{n}(\widehat{\sigma}_n^2 - \sigma^2) < x\}\right| \to 0 \tag{2.3.2}$$

where P^* *denotes the conditional probability given* $\mathbf{Y}$.

Recalling the decompositions for $\sqrt{n}(\beta_{LS} - \beta)$ and $\sqrt{n}(\widehat{\sigma}_n^2 - \sigma^2)$ in (5.1.3) and (5.1.4) below, and applying these to$\sqrt{n}(\beta_{LS}^* - \beta_{LS})$ and $\sqrt{n}(\widehat{\sigma}_n^{2*} - \widehat{\sigma}_n^2)$, we can calculate the tail probability value of each term explicitly. The proof is similar to those given in Sections 5.1 and 5.2. We refer the details to Liang, Härdle and Sommerfeld (1999).

We have now shown that the bootstrap method performs at least as good as the normal approximation with the error rate of $o_p(1)$. It is natural to expect that the bootstrap method should perform better than this in practice. As a matter of fact, our numerical results support this conclusion. Furthermore, we have the following theoretical result. Let $M_{jn}(\beta)$ $[M_{jn}(\sigma^2)]$ and $M_{jn}^*(\beta)$ $[M_{jn}^*(\sigma^2)]$ be the j−th moments of $\sqrt{n}(\beta_{LS} - \beta)$ $[\sqrt{n}(\widehat{\sigma}_n^2 - \sigma^2)]$ and $\sqrt{n}(\beta_{LS}^* - \beta_{LS})$ $[\sqrt{n}(\widehat{\sigma}_n^{2*} - \widehat{\sigma}_n^2)]$, respectively.

Theorem 2.3.2 *Assume that Assumptions 1.3.1-1.3.3 hold. Let* $E\varepsilon_1^6 < \infty$ *and* $\max_{1\leq i\leq n}\|u_i\| \leq C_0 < \infty$. *Then* $M_{jn}^*(\beta) - M_{jn}(\beta) = O_P(n^{-1/3}\log n)$ *and* $M_{jn}^*(\sigma^2) - M_{jn}(\sigma^2) = O_P(n^{-1/3}\log n)$ *for* $j = 1, 2, 3, 4$.

The proof of Theorem 2.3.2 follows similarly from that of Theorem 2.3.1. We omit the details here.

Theorem 2.3.2 shows that the bootstrap distributions have much better approximations for the first four moments of β^*_{LS} and $\widehat{\sigma}^{2*}_n$. The first four moments are the most important quantities in characterizing distributions. In fact, Theorems 2.3.1 and 2.1.1 can only obtain

$$M^*_{jn}(\beta) - M_{jn}(\beta) = o_P(1) \quad \text{and } M^*_{jn}(\sigma^2) - M_{jn}(\sigma^2) = o_P(1)$$

for $j = 1, 2, 3, 4$.

2.3.3 Numerical Results

In this subsection we present a small simulation study to illustrate the finite sample behaviors of the estimators. We investigate the model given by

$$Y_i = X_i^T\beta + g(T_i) + \varepsilon_i \tag{2.3.3}$$

where $g(T_i) = \sin(T_i)$, $\beta = (1,5)'$ and $\varepsilon_i \sim$ Uniform$(-0.3, 0.3)$. The mutually independent variables $X_i = (X_i^{(1)}, X_i^{(2)})$ and T_i are realizations of a Uniform$(0,1)$ distributed random variable. We analyze (2.3.3) for the sample sizes of 30, 50, 100 and 300. For nonparametric fitting, we use a `Nadaraya-Watson kernel weight` function with `Epanechnikov kernel`. We perform the smoothing with different bandwidths using some grid search. Our simulations show that the results for the parametric part are quite robust against the bandwidth chosen in the nonparametric part. In the following we present only the simulation results for the parameter β_2. Those for β_1 are similar.

We implement our small sample studies for the cases of sample sizes 30, 50, 100, 300. In Figure 2.3, we plot the smoothed densities of the estimated true distribution of $\sqrt{n}(\widehat{\beta}_2 - \beta_2)/\widehat{\sigma}$ with $\widehat{\sigma}^2 = 1/n\sum_{i=1}^n(\tilde{Y}_i - \tilde{X}_i^T\beta_n)^2$. Additionally we depict the corresponding bootstrap distributions and the asymptotic normal distributions, in which we estimate σ^2B^{-1} by $\widehat{\sigma}^2\widehat{B}^{-1}$ with $\widehat{B} = 1/n\sum_{i=1}^n \widetilde{X}_i\widetilde{X}_i^T$. It turns out that the bootstrap distribution and the asymptotic normal distribution approximate the true ones very well even when the sample size of n is only 30.

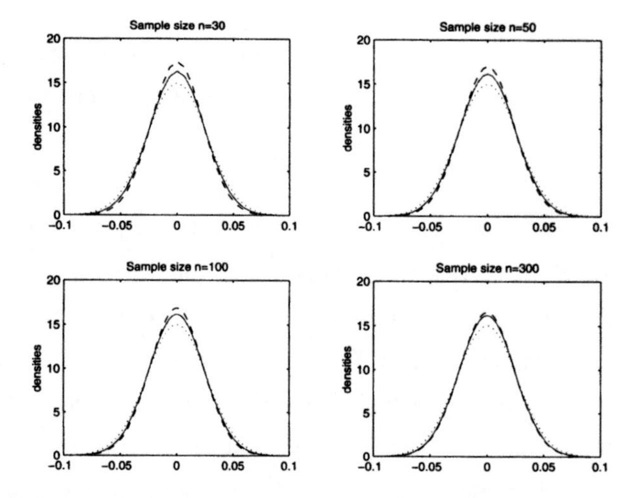

FIGURE 2.3. Plot of of the smoothed bootstrap density (dashed), the normal approximation (dotted) and the smoothed true density (solid).

3 ESTIMATION OF THE NONPARAMETRIC COMPONENT

3.1 Introduction

In this chapter, we will focus on deriving the asymptotic properties of an estimator of the unknown function $g(\cdot)$ for the case of `fixed design points`. We consider its consistency, `weak convergence` rate and asymptotic normality. We also derive these results for a specific version of (1.2.3) with nonstochastic regressors and `heteroscedastic` errors.

Previous work in a `heteroscedastic` setting has focused on the nonparametric regression model (i.e. $\beta = 0$). Müller and Stadtmüller (1987) proposed an estimate of the variance function by using a kernel smoother, and then proved that the estimate is uniformly consistent. Hall and Carroll (1989) considered the consistency of estimates of $g(\cdot)$. Eubank and Whitney (1989) proposed `trigonometric series` type estimators of g. They investigated asymptotic approximations of the integrated mean squared error and the partial integrated mean squared error of g_λ.

Well-known applications in econometrics literature that can be put in the form of (1.1.1) are the `human capital` earnings function (Willis (1986)) and the `wage curve` (Blanchflower and Oswald, 1994). In both cases, log-earnings of an individual are related to personal characteristics (sex, marital status) and measures of a person's `human capital` like schooling and labor market experience. Economic theory suggests a non-linear relationship between log-earnings and labor market experience. The `wage curve` is obtained by including the `local unemployment rate` as an additional regressor, with a possibly non-linear influence. Rendtel and Schwarze (1995), for instance, estimated $g(\cdot)$ as a function of the `local unemployment rate` using smoothing-splines and found a `U-shaped` relationship.

The organization of this chapter is as follows. Weak and strong consistency results are given in Section 3.2. The asymptotic normality of the nonparametric estimator of g is given in Section 3.3. In Section 3.4, we illustrate our estimation procedure by a small-scale Monte Carlo study and an empirical illustration.

3.2 Consistency Results

In order to establish some consistency results, we introduce the following assumptions.

Assumption 3.2.1 *Assume that the following equations hold uniformly over* $[0,1]$ *and* $n \geq 1$*:*

(a) $\sum_{i=1}^{n} |\omega_{ni}(t)| \leq C_1$ *for all* t *and some constant* C_1*;*

(b) $\sum_{i=1}^{n} \omega_{ni}(t) - 1 = O(\mu_n)$ *for some* $\mu_n > 0$*;*

(c) $\sum_{i=1}^{n} |\omega_{ni}(t)| I(|t - T_i| > \mu_n) = O(\mu_n)$*;*

(d) $\sup_{i \leq n} |\omega_{ni}(t)| = O(\nu_n^{-1})$*;*

(e) $\sum_{i=1}^{n} \omega_{ni}^2(t) E\varepsilon_i^2 = \sigma_0^2/\nu_n + o(1/\nu_n)$ *for some* $\sigma_0^2 > 0$.

where both μ_n *and* ν_n *are positive sequences satisfying* $\lim_{n\to\infty} \mu_n = 0$, $\lim_{n\to\infty} \nu_n/n = 0$, $\lim_{n\to\infty} n^{1/2}\log n/\nu_n = 0$, *and* $\limsup_{n\to\infty} \mu_n \nu_n^2 < \infty$.

Assumption 3.2.2 *The weight functions* ω_{ni} *satisfy*

$$\max_{i\geq 1} |\omega_{ni}(s) - \omega_{ni}(t)| \leq C_2 |s - t|$$

uniformly over $n \geq 1$ *and* $s, t \in [0,1]$*, where* C_2 *is a bounded constant.*

We now have the following result.

Theorem 3.2.1 *Assume that Assumptions 1.3.1 and 3.2.1 hold. Then at every continuity point of the function* g*,*

$$E\{\widehat{g}_n(t) - g(t)\}^2 = \frac{\sigma_0^2}{\nu_n} + O(\mu_n^2) + o(\nu_n^{-1}) + o(\mu_n^2).$$

Proof. Observe the following decomposition:

$$\widehat{g}_n(t) - g(t) = \sum_{j=1}^{n} \omega_{nj}(t)\varepsilon_j + \sum_{j=1}^{n} \omega_{nj}(t) X_j^T(\beta - \beta_{LS}) + \sum_{j=1}^{n} \omega_{nj}(t) g(T_j) - g(t). \tag{3.2.1}$$

Similar to Lemma A.1, we have

$$\sum_{j=1}^{n} \omega_{nj}(t) g(T_j) - g(t) = O(\mu_n)$$

at every continuity point of g.

Similar to the proof of Lemma A.2, we have for $1 \le j \le p$,

$$\begin{aligned}
\Big|\sum_{i=1}^{n}\omega_{ni}(t)x_{ij}\Big| &= \Big|\sum_{i=1}^{n}\omega_{ni}(t)\{u_{ij}+h_j(T_i)\}\Big| \\
&= O(1)+\Big|\sum_{i=1}^{n}\omega_{ni}(t)u_{ij}\Big| \\
&\le O(1)+6\sup_{1\le i\le n}|\omega_{ni}(t)|\max_{1\le k\le n}\Big|\sum_{i=1}^{k}u_{j_i m}\Big| \\
&= O(1)+O(n^{1/2}\log n\nu_n^{-1})=O(1). \qquad (3.2.2)
\end{aligned}$$

On the other hand, it follows from the definition of β_{LS} that

$$E(\beta_{LS}-\beta)^2=E(\beta_{LS}-E\beta_{LS})^2+(E\beta_{LS}-\beta)^2=O(n^{-1}), \qquad (3.2.3)$$

which is a direct calculation.

Similarly, we have

$$E\Big\{\sum_{i=1}^{n}\omega_{ni}(t)\varepsilon_i\Big\}^2=\sum_{i=1}^{n}\omega_{ni}^2(t)E\varepsilon_i^2=\frac{\sigma_0^2}{\nu_n}+o(\nu_n^{-1}) \qquad (3.2.4)$$

using Assumptions 3.2.1 (e).

Therefore, (3.2.1)-(3.2.4) imply

$$E\{\widehat{g}_n(t)-g(t)\}^2=\frac{\sigma_0^2}{\nu_n}+O(\mu_n^2)+o(\nu_n^{-1})+o(\mu_n^2). \qquad (3.2.5)$$

Remark 3.2.1 *Equation (3.2.5) not only provides an optimum convergence rate, but proposes a theoretical selection for the smoothing parameter involved in the weight functions ω_{ni} as well. For example, when considering $\omega_{ni}^{(1)}$ or $\omega_{ni}^{(2)}$ defined in Remark 1.3.1 as ω_{ni}, $\nu_n=nh_n$, and $\mu_n=h_n^2$, under Assumptions 1.3.1, 3.2.1 and 3.2.2, we have*

$$E\{\widehat{g}_n(t)-g(t)\}^2=\frac{\sigma_0^2}{nh_n}+O(h_n^4)+o(n^{-1}h_n^{-1})+o(h_n^4). \qquad (3.2.6)$$

This suggests that the theoretical selection of h_n is proportional to $n^{-1/5}$. Details about the practical selection of h_n are similar to those in Section 6.4.

Remark 3.2.2 *In the proof of (3.2.2), we can assume that for all $1 \le j \le p$,*

$$\sum_{i=1}^{n}\omega_{ni}(t)u_{ij}=O(s_n) \qquad (3.2.7)$$

uniformly over $t \in [0,1]$, *where* $s_n \to 0$ *as* $n \to \infty$. *Since the real sequences* u_{ij} *behave like i.i.d. random variables with mean zero, equation (3.2.7) is reasonable. Actually, both Assumption 1.3.1 and equation (3.2.7) hold with probability one when* (X_i, T_i) *are independent random designs.*

We now establish the strong rate of convergence of $\widehat{g}_n$.

Theorem 3.2.2 *Assume that Assumptions 1.3.1, 1.3.2, 3.2.1 and 3.2.2 hold. Let* $E|\varepsilon_1|^3 < \infty$. *Then*

$$\sup_t |\widehat{g}_n(t) - g(t)| = O_P(\nu_n^{-1/2} \log^{1/2} n) + O(\mu_n) + O_P(n^{-1/2}). \qquad (3.2.8)$$

Proof. It follows from Assumption 3.2.1 that

$$\sup_t \Big| \sum_{j=1}^n \omega_{nj}(t) g(T_j) - g(t) \Big| = O(\mu_n). \qquad (3.2.9)$$

Similar to the proof of Lemma 5.1 of Müller and Stadtmüller (1987), we have

$$\sup_t \Big| \sum_{j=1}^n \omega_{nj}(t) \varepsilon_i \Big| = O_P(\nu_n^{-1} \log^{1/2} n). \qquad (3.2.10)$$

The details are similar to those of Lemma A.3 below and have been given in Gao, Chen and Zhao (1994). Therefore, the proof of (3.2.8) follows from (3.2.1), (3.2.2), (3.2.9), (3.2.10) and the fact $\beta_{LS} - \beta = O_P(n^{-1/2})$.

Remark 3.2.3 *Theorem 3.2.2 shows that the estimator of the nonparametric component in (1.1.1) can achieve the optimum rate of convergence for the completely nonparametric regression.*

Theorem 3.2.3 *Assume that Assumptions 1.3.1, 3.2.1 and 3.2.2 hold. Then* $\nu_n Var\{\widehat{g}_n(t)\} \to \sigma_0^2$ *as* $n \to \infty$.

Proof.

$$\begin{aligned} \nu_n Var\{\widehat{g}_n(t)\} &= \nu_n E\Big\{\sum_{i=1}^n \omega_{ni}(t)\varepsilon_i\Big\}^2 + \nu_n E\Big\{\sum_{i=1}^n \omega_{ni}(t) X_i^T (\widetilde{\mathbf{X}}^T \widetilde{\mathbf{X}})^{-1} \widetilde{\mathbf{X}}^T \widetilde{\varepsilon}\Big\}^2 \\ &\quad -2\nu_n E\Big\{\sum_{i=1}^n \omega_{ni}(t)\varepsilon_i\Big\} \cdot \Big\{\sum_{i=1}^n \omega_{ni}(t) X_i^T (\widetilde{\mathbf{X}}^T \widetilde{\mathbf{X}})^{-1} \widetilde{\mathbf{X}}^T \widetilde{\varepsilon}\Big\}. \end{aligned}$$

The first term converges to σ_0^2. The second term tends to zero, since

$$E\Big\{\sum_{i=1}^n \omega_{ni}(t) X_i^T (\widetilde{\mathbf{X}}^T \widetilde{\mathbf{X}})^{-1} \widetilde{\mathbf{X}}^T \widetilde{\varepsilon}\Big\}^2 = O(n^{-1}).$$

Using `Cauchy-Schwarz inequality`, the third term is shown to tend to zero.

3.3 Asymptotic Normality

In the nonparametric regression model, Liang (1995c) proved asymptotic normality for independent ε_i's under some mild conditions. In this section, we shall consider the asymptotic normality of $\widehat{g}_n(t)$ under the Assumptions 1.3.1, 3.2.1 and 3.2.2.

Theorem 3.3.1 *Assume that $\varepsilon_1, \varepsilon_2, \ldots, \varepsilon_n$ are independent random variables with $E\varepsilon_i = 0$ and $\inf_i E\varepsilon_i^2 > c_\sigma > 0$ for some c_σ. There exists a function $G(u)$ satisfying*

$$\int_0^\infty uG(u)du < \infty \tag{3.3.1}$$

such that

$$P(|\varepsilon_i| > u) \le G(u), \quad \text{for } i = 1, \ldots, n \quad \text{and large enough } u. \tag{3.3.2}$$

If

$$\frac{\max_{1\le i\le n} \omega_{ni}^2(t)}{\sum_{i=1}^n \omega_{ni}^2(t)} \to 0 \qquad \text{as} \quad n \to \infty, \tag{3.3.3}$$

then

$$\frac{\widehat{g}_n(t) - E\widehat{g}_n(t)}{\sqrt{Var\{\widehat{g}_n(t)\}}} \longrightarrow^{\mathcal{L}} N(0,1) \qquad \text{as } n \to \infty.$$

Remark 3.3.1 *The conditions (3.3.1) and (3.3.2) guarantee* $\sup_i E\varepsilon_i^2 < \infty$.

The proof of Theorem 3.3.1. It follow from the proof of Theorem 3.2.3 that

$$Var\{\widehat{g}_n(t)\} = \sum_{i=1}^n \omega_{ni}^2(t)\sigma_i^2 + o\Big\{\sum_{i=1}^n \omega_{ni}^2(t)\sigma_i^2\Big\}.$$

Furthermore

$$\widehat{g}_n(t) - E\widehat{g}_n(t) - \sum_{i=1}^n \omega_{ni}(t)\varepsilon_i = \sum_{i=1}^n \omega_{ni}(t)X_i^T(\widetilde{\mathbf{X}}^T\widetilde{\mathbf{X}})^{-1}\widetilde{\mathbf{X}}^T\widetilde{\varepsilon} = O_P(n^{-1/2}),$$

which yields

$$\frac{\sum_{i=1}^n \omega_{ni}(t)X_i^T(\widetilde{\mathbf{X}}^T\widetilde{X})^{-1}\widetilde{\mathbf{X}}^T\widetilde{\varepsilon}}{\sqrt{Var\{\widehat{g}_n(t)\}}} = O_P(n^{-1/2}\nu_n^{1/2}) = o_P(1).$$

It follows that

$$\frac{\widehat{g}_n(t) - E\widehat{g}_n(t)}{\sqrt{Var\{\widehat{g}_n(t)\}}} = \frac{\sum_{i=1}^n \omega_{ni}(t)\varepsilon_i}{\sqrt{\sum_{i=1}^n \omega_{ni}^2(t)\sigma_i^2}} + o_P(1) \stackrel{\text{def}}{=} \sum_{i=1}^n a_{ni}^*\varepsilon_i + o_P(1),$$

where $a_{ni}^* = \frac{\omega_{ni}(t)}{\sqrt{\sum_{i=1}^n \omega_{ni}^2(t)\sigma_i^2}}$. Let $a_{ni} = a_{ni}^*\sigma_i$. Obviously, this means that $\sup_i \sigma_i < \infty$, due to $\int_0^\infty vG(v)dv < \infty$. The proof of the theorem immediately follows from (3.3.1)–(3.3.3) and Lemma 3.5.3 below.

Remark 3.3.2 *If $\varepsilon_1, \ldots, \varepsilon_n$ are i.i.d., then $E|\varepsilon_1|^2 < \infty$ and the condition (3.3.3) of Theorem 3.3.1 can yield the result of Theorem 3.3.1.*

Remark 3.3.3 *(a) Let ω_{ni} be either $\omega_{ni}^{(1)}$ or $\omega_{ni}^{(2)}$ defined in Remark 1.3.1, $\nu_n = nh_n$, and $\mu_n = h_n^2$. Assume that the probability kernel function K satisfies: (i) K has compact support; (ii) the first two derivatives of K are bounded on the compact support of K. Then Theorem 3.3.1 implies that as $n \to \infty$,*

$$\sqrt{nh_n}\{\widehat{g}_n(t) - E\widehat{g}_n(t)\} \longrightarrow^{\mathcal{L}} N(0, \sigma_0^2).$$

This is the classical conclusion in nonparametric regression. See Härdle (1990).

(b) If we replace the condition $\limsup_{n\to\infty} \nu_n\mu_n^2 < \infty$ by $\lim_{n\to\infty} \nu_n\mu_n^2 = 0$ in Assumption 3.2.1, then as $n \to \infty$,

$$\frac{\widehat{g}_n(t) - g(t)}{\sqrt{Var\{\widehat{g}_n(t)\}}} \longrightarrow^{\mathcal{L}} N(0,1).$$

Obviously, the selection of bandwidth in the kernel regression case is not asymptotically optimal since the bandwidth satisfies $\lim_{n\to\infty} nh_n^5 = 0$. In general, asymptotically optimal estimate in the kernel regression case always has a nontrivial bias. See Härdle (1990).

3.4 Simulated and Real Examples

In this section, we illustrate the finite-sample behavior of the estimator by applying it to real data and by performing a small simulation study.

Example 3.4.1 *In the introduction, we mentioned the* `human capital` *earnings function as a well-known econometric application that can be put into the form of a partially linear model. It typically relates the logarithm of earnings to a set of explanatory variables describing an individual's skills, personal characteristics and labour market conditions. Specifically, we estimate β and $g(\cdot)$ in the model*

$$\ln Y_i = X_i^T\beta + g(T_i) + \varepsilon_i, \tag{3.4.1}$$

where X contains two dummy variables indicating that the level of secondary schooling a person has completed, and T is a measure of labour market experience (defined as the number of years spent in the labour market and approximated by subtracting (years of schooling + 6) from a person's age).

Under certain assumptions, the estimate of β can be interpreted as the rate of return from obtaining the respective level of secondary schooling. Regarding $g(T)$, `human capital` theory suggests a concave form: Rapid `human capital` accumulation in the early stage of one's labor market career is associated with rising earnings that peak somewhere during midlife and decline thereafter as hours worked and the incentive to invest in `human capital` decreases. To allow for concavity, parametric specifications of the earnings-function typically include T and T^2 in the model and obtain a positive estimate for the coefficient of T and a negative estimate for the coefficient of T^2.

For nonparametric fitting, we use a `Nadaraya-Watson weight function` with quartic kernel

$$(15/16)(1-u^2)^2 I(|u| \leq 1)$$

and choose the `bandwidth` using `cross-validation`. The estimate of $g(T)$ is given in Figure 3.1. When a sample size is smaller than that used in most empirical investigations of the `human capital` earnings function, we obtain a nonparametric estimate that nicely agrees with the concave relationship envisioned by economic theory.

Remark 3.4.1 *Figure 3.1 shows that the relation between predicted earnings and the level of experience is nonlinear. This conclusion is the same as that reached by using the classical parametric fitting. Empirical economics suggests using a second-order polynomial to fit the relationship between the predicted earnings and the level experience. Our nonparametric approach provides a better fitting between the two.*

Remark 3.4.2 *The above Example 3.4.1 demonstrates that partially linear regression is better than the classical linear regression for fitting some economic data. Recently, Anglin and Gencay (1996) considered another application of partially linear regression to a hedonic price function. They estimated a benchmark parametric model which passes several common specification tests before showing*

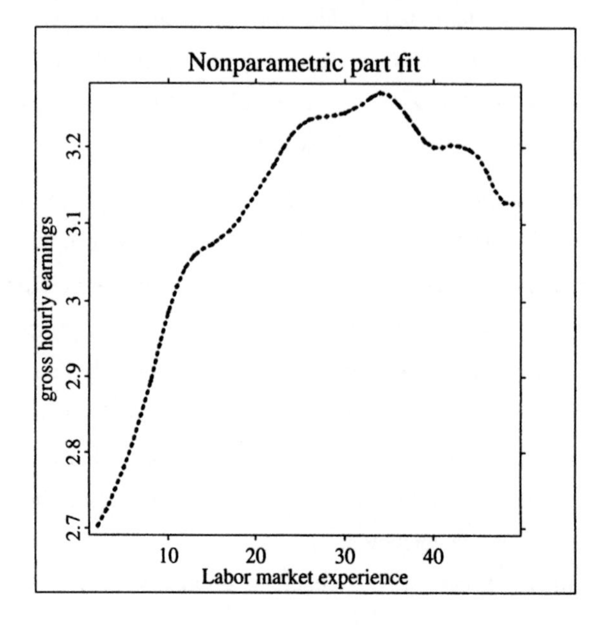

FIGURE 3.1. Relationship between log-earnings and labour-market experience

that a partially linear model outperforms it significantly. Their research suggests that the partially linear model provides more accurate mean predictions than the benchmark parametric model. See also Liang and Huang (1996), who discussed some applications of partially linear regression to economic data.

Example 3.4.2 *We also conduct a small simulation study to get further small-sample properties of the estimator of $g(\cdot)$. We consider the model*

$$Y_i = X_i^T\beta + \sin(\pi T_i) + \sin(X_i^T\beta + T_i)\xi_i, \qquad i = 1, \ldots, n = 300$$

where $\{\xi_i\}$ is sequence of i.i.d. standard normal errors, $X_i = (X_{i1}, X_{i2})^T$, and $X_{i1} = X_{i2} = T_i = i/n$. We set $\beta = (1, 0.75)^T$ and perform 100 `replications` *of generating samples of size $n = 300$. Figure 3.2 presents the "true" curve $g(T) = \sin(\pi T)$ (solid-line) and an average of the 100 estimates of $g(\cdot)$ (dashed-line). The average estimate nicely captures the shape of $g(\cdot)$.*

3.5 Appendix

In this section we state some useful lemmas.

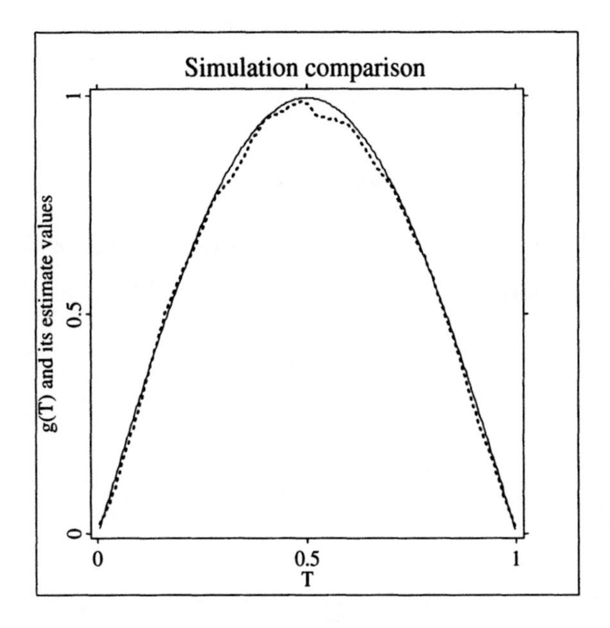

FIGURE 3.2. Estimates of the function $g(T)$.

Lemma 3.5.1 *Suppose that Assumptions 3.2.1 and 3.2.2 hold and that $g(\cdot)$ and $h_j(\cdot)$ are continuous. Then*

$$(i) \qquad \max_{1\le i\le n}\Big|G_j(T_i) - \sum_{k=1}^{n}\omega_{nk}(T_i)G_j(T_k)\Big| = o(1).$$

Furthermore, suppose that $g(\cdot)$ and $h_j(\cdot)$ are `Lipschitz continuous` *of order 1. Then*

$$(ii) \qquad \max_{1\le i\le n}\Big|G_j(T_i) - \sum_{k=1}^{n}\omega_{nk}(T_i)G_j(T_k)\Big| = O(c_n)$$

for $j = 0,\dots,p$, where $G_0(\cdot) = g(\cdot)$ and $G_l(\cdot) = h_l(\cdot)$ for $l = 1,\dots,p$.

Proof. The proofs are similar to Lemma A.1. We omit the details.

The following Lemma is a slightly modified version of Theorem 9.1.1 of Chow and Teicher (1988). We therefore do not give its proof.

Lemma 3.5.2 *Let $\xi_{nk}, k = 1,\dots,k_n$, be independent random variables with mean zero and finite variance σ_{nk}^2. Assume that $\lim_{n\to\infty}\sum_{k=1}^{k_n}\sigma_{nk}^2 = 1$ and $\max_{1\le k\le k_n}\sigma_{nk}^2 \to 0$. Then $\sum_{k=1}^{k_n}\xi_{nk} \to^{\mathcal{L}} N(0,1)$ if and only if*

$$\sum_{k=1}^{k_n} E\xi_{nk}^2 I(|\xi_{nk}| > \delta) \to 0 \quad \textit{for any } \delta > 0 \quad \textit{as } n \to \infty.$$

Lemma 3.5.3 *Let $V_1, \ldots, V_n$ be independent random variables with $EV_i = 0$ and $\inf_i EV_i^2 > C > 0$ for some constant number C. There exists a function $H(v)$ satisfying $\int_0^\infty vH(v)dv < \infty$ such that*

$$P\{|V_k| > v\} \le H(v) \quad \textit{for large enough } v > 0 \quad \textit{and } k = 1, \ldots, n. \tag{3.5.1}$$

Also assume that $\{a_{ni}, i = 1, \ldots, n, n \ge 1\}$ is a sequence of real numbers satisfying $\sum_{i=1}^n a_{ni}^2 = 1$. If $\max_{1\le i\le n} |a_{ni}| \to 0$, then for $a'_{ni} = a_{ni}/\sigma_i(V)$,

$$\sum_{i=1}^n a'_{ni} V_i \longrightarrow^{\mathcal{L}} N(0,1) \quad \textit{as } n \to \infty.$$

where $\{\sigma_i(V)\}$ is the variance of $\{V_i\}$.

Proof. Denote $\xi_{nk} = a'_{nk} V_k, k = 1, \ldots, n$. We have $\sum_{k=1}^n E\xi_{nk}^2 = 1$. Moreover, it follows that

$$\begin{aligned}
\sum_{k=1}^n E\{\xi_{nk}^2 I(|\xi_{nk}| > \delta)\} &= \sum_{k=1}^n {a'_{nk}}^2 E\{V_k^2 I(|a_{nk}V_k| > \delta)\} \\
&= \sum_{k=1}^n \frac{a_{nk}^2}{\sigma_k^2} E\{V_k^2 I(|a_{nk}V_k| > \delta)\} \\
&\le (\inf_k \sigma_k^2)^{-1} \sup_k E\{V_k^2 I(|a_{nk}V_k| > \delta)\}.
\end{aligned}$$

It follows from the condition (3.5.1) that

$$\sup_k E\{V_k^2 I(|a_{nk}V_k| > \delta)\} \to 0 \quad \text{as } n \to \infty.$$

Therefore Lemma 3.5.3 follows from Lemma 3.5.2.

4 ESTIMATION WITH MEASUREMENT ERRORS

4.1 Linear Variables with Measurement Errors

4.1.1 Introduction and Motivation

In this section, we are interested in the estimation of the unknown parameter β and the unknown function $g(\cdot)$ in model (1.1.1) when the covariates X_i are measured with errors. Instead of observing X_i, we observe

$$W_i = X_i + U_i, \tag{4.1.1}$$

where the **measurement errors** U_i are i.i.d., independent of (Y_i, X_i, T_i), with mean zero and covariance matrix Σ_{uu}. We will assume that Σ_{uu} is known, taking up the case that it is estimated in subsection 4.1.4. The measurement error literature has been surveyed by Fuller (1987) and Carroll, Ruppert and Stefanski (1995).

It is well known that in linear regression, by applying the so–called **correction for attenuation**, inconsistency caused by measurement error can be overcome. In our context, this suggests that we use the estimator

$$\widehat{\beta}_n = (\widetilde{\mathbf{W}}^T\widetilde{\mathbf{W}} - n\Sigma_{uu})^{-1}\widetilde{\mathbf{W}}^T\widetilde{\mathbf{Y}}. \tag{4.1.2}$$

The estimator (4.1.2) can be derived in much the same way as the Severini–Staniswalis estimator. For every β, let $\widehat{g}(T, \beta)$ maximize the weighted likelihood ignoring **measurement error**, and then form an estimator of β via a negatively penalized operation:

$$\text{minimize} \sum_{i=1}^{n} \left\{Y_i - W_i^T\beta - \widehat{g}(T_i, \beta)\right\}^2 - \beta^T\Sigma_{uu}\beta. \tag{4.1.3}$$

The fact that $g(t) = E(Y_i - W_i^T\beta | T = t)$ suggests

$$\widehat{g}_{nw}(t) = \sum_{j=1}^{n} \omega_{nj}(t)(Y_j - W_j^T\widehat{\beta}_n) \tag{4.1.4}$$

as the estimator of $g(t)$.

In some cases, it may be reasonable to assume that the model errors ε_i are homoscedastic with common variance σ^2. In this event, since $E\{Y_i - X_i^T\beta - g(T_i)\}^2 = \sigma^2$ and $E\{Y_i - W_i^T\beta - g(T_i)\}^2 = E\{Y_i - X_i^T\beta - g(T_i)\}^2 + \beta^T\Sigma_{uu}\beta$, we define

$$\widehat{\sigma}_n^2 = n^{-1}\sum_{i=1}^{n}(\widetilde{Y}_i - \widetilde{W}_i^T\widehat{\beta}_n)^2 - \widehat{\beta}_n^T\Sigma_{uu}\widehat{\beta}_n \tag{4.1.5}$$

as the estimator of σ^2. The negative sign in the second term in (4.1.3) looks odd until one remembers that the effect of measurement error is attenuation, i.e., to underestimate β in absolute value when it is scalar, and thus one must correct for attenuation by making β larger, not by shrinking it further towards zero.

In this chapter, we analyze the estimate (4.1.2) and show that it is consistent, asymptotically normally distributed with a variance different from (2.1.1). Just as in the Severini–Staniswalis algorithm, a kernel weighting ordinary bandwidth of order $h \sim n^{-1/5}$ may be used.

Subsection 4.1.2 is the statement of the main results for β, while the results for $g(\cdot)$ are stated in Subsection 4.1.3. Subsection 4.1.4 states the corresponding results for the measurement error variance Σ_{uu} estimated. Subsection 4.1.5 gives a numerical illustration. Several remarks are given in Subsection 4.1.6. All proofs are delayed until the last subsection.

4.1.2 Asymptotic Normality for the Parameters

Our two main results are concerned with the limit distributions of the estimates of β and σ^2.

Theorem 4.1.1 *Suppose that Assumptions 1.3.1-1.3.3 hold and that* $E(\varepsilon^4 + \|U\|^4) < \infty$. *Then* $\widehat{\beta}_n$ *is an asymptotically normal estimator, i.e.*

$$n^{1/2}(\widehat{\beta}_n - \beta) \longrightarrow^{\mathcal{L}} N(0, \Sigma^{-1}\Gamma\Sigma^{-1}),$$

where $\Gamma = E[(\varepsilon - U^T\beta)\{X - E(X|T)\}]^{\otimes 2} + E\{(UU^T - \Sigma_{uu})\beta\}^{\otimes 2} + E(UU^T\varepsilon^2)$. *Note that* $\Gamma = E(\varepsilon - U^T\beta)^2\Sigma + E\{(UU^T - \Sigma_{uu})\beta\}^{\otimes 2} + \Sigma_{uu}\sigma^2$ *if* ε *is* homoscedastic *and independent of* (X, T).

Theorem 4.1.2 *Suppose that the conditions of Theorem 4.1.1 hold. In addition, we assume that the* ε*'s are* homoscedastic *with variance* σ^2, *and independent of*

(X, T). *Then*

$$n^{1/2}(\widehat{\sigma}_n^2 - \sigma^2) \longrightarrow^{\mathcal{L}} N(0, \sigma_*^2),$$

where $\sigma_*^2 = E\{(\varepsilon - U^T\beta)^2 - (\beta^T\Sigma_{uu}\beta + \sigma^2)\}^2$.

Remarks

- As described in the introduction, an important aspect of the results of Severini and Staniswalis is that their methods lead to asymptotically normal parameter estimates in kernel regression, even with the bandwidth of the usual order $h_n \approx n^{-1/5}$. The same holds for our estimators in general. For example, suppose that the design points T_i satisfy that there exist constants M_1 and M_2 such that

  $$M_1/n \le \min_{i \le n} |T_i - T_{i-1}| \le \max_{i \le n} |T_i - T_{i-1}| \le M_2/n.$$

  Then Assumptions 1.3.3(i)-(iii) are satisfied by a simple verification.

- It is relatively easy to estimate the covariance matrix of $\widehat{\beta}_n$. Let $\dim(X)$ be the number of the components of X. A consistent estimate of Σ is just

  $$\{n - \dim(X)\}^{-1} \sum_{i=1}^{n} \{W_i - \widehat{g}_{w,h}(T_i)\}^{\otimes 2} - \Sigma_{uu} (\stackrel{\text{def}}{=} \widehat{\Sigma}_n).$$

  In the general case, one can use (4.1.15) to construct a consistent sandwich--type estimate of Γ, namely

  $$n^{-1} \sum_{i=1}^{n} \left\{\widetilde{W}_i(\widetilde{Y}_i - \widetilde{W}_i^T\widehat{\beta}_n) + \Sigma_{uu}\widehat{\beta}_n\right\}^{\otimes 2}.$$

  In the homoscedastic case, namely that ε is independent of (X, T, U) with variance σ^2, and with U being normally distributed, a different formula can be used. Let $\mathcal{C}(\beta) = E\{(UU^T - \Sigma_{uu})\beta\}^{\otimes 2}$. Then a consistent estimate of Γ is

  $$(\widehat{\sigma}_n^2 + \widehat{\beta}_n^T\Sigma_{uu}\widehat{\beta}_n)\widehat{\Sigma}_n + \widehat{\sigma}_n^2\Sigma_{uu} + \mathcal{C}(\widehat{\beta}_n).$$

- In the classical functional model, instead of obtaining an estimate of Σ_{uu} through replication, it is instead assumed that the ratio of Σ_{uu} to σ^2 is known. Without loss of generality, we set this ratio equal to the identity matrix. The resulting analogue of the parametric estimators to the partially linear model is to solve the following minimization problem:

$$\sum_{i=1}^{n}\left|\frac{\widetilde{Y}_i-\widetilde{W}_i^T\beta}{\sqrt{1+\|\beta\|^2}}\right|^2=\min!,$$

where $\|\cdot\|$ denotes the Euclidean norm. One can use the techniques of this section to show that this estimator is consistent and asymptotically normally distributed. The asymptotic variance of the estimate of β for the case where ε is independent of (X,T) can be shown to be

$$\Sigma^{-1}\left[(1+\|\beta\|^2)^2\sigma^2\Sigma+\frac{E\{(\varepsilon-U^T\beta)^2\Gamma_1\Gamma_1^T\}}{1+\|\beta\|^2}\right]\Sigma^{-1},$$

where $\Gamma_1=(1+\|\beta\|^2)U+(\varepsilon-U^T\beta)\beta$.

4.1.3 Asymptotic Results for the Nonparametric Part

Theorem 4.1.3 *Suppose that Assumptions 1.3.1-1.3.3 hold and that $\omega_{ni}(t)$ are* `Lipschitz continuous` *of order 1 for all $i=1,\dots,n$. If $E(\varepsilon^4+\|U\|^4)<\infty$, then for fixed T_i, the asymptotic bias and asymptotic variance of $\widehat{g}_{nw}(t)$ are $\sum_{i=1}^n\omega_{ni}(t)g(T_i)-g(t)$ and $\sum_{i=1}^n\omega_{ni}^2(t)(\beta^T\Sigma_{uu}\beta+\sigma^2)$, respectively.*

If (X_i,T_i) are random, then the bias and variance formulas are the usual ones for nonparametric kernel regression.

4.1.4 Estimation of Error Variance

Although in some cases, the `measurement error` covariance matrix Σ_{uu} has been established by independent experiments, in others it is unknown and must be estimated. The usual method of doing so is by partial `replication`, so that we observe $W_{ij}=X_i+U_{ij},\quad j=1,...m_i$.

We consider here only the usual case that $m_i\le 2$, and assume that a fraction δ of the data has such replicates. Let $\overline{W}_i$ be the sample mean of the replicates. Then a consistent, unbiased method of moments estimate for Σ_{uu} is

$$\widehat{\Sigma}_{uu}=\frac{\sum_{i=1}^n\sum_{j=1}^{m_i}(W_{ij}-\overline{W}_i)^{\otimes 2}}{\sum_{i=1}^n(m_i-1)}.$$

The estimator changes only slightly to accommodate the replicates, becoming

$$\begin{aligned}\widehat{\beta}_n &= \left[\sum_{i=1}^n\left\{\overline{W}_i-\widehat{g}_{w,h}(T_i)\right\}^{\otimes 2}-n(1-\delta/2)\widehat{\Sigma}_{uu}\right]^{-1}\\ &\quad\times\sum_{i=1}^n\left\{\overline{W}_i-\widehat{g}_{w,h}(T_i)\right\}\left\{Y_i-\widehat{g}_{y,h}(T_i)\right\},\end{aligned}\tag{4.1.6}$$

where $\widehat{g}_{w,h}(\cdot)$ is the kernel regression of the $\overline{W}_i$'s on T_i.

Using the techniques in Subsection 4.1.7, one can show that the limit distribution of (4.1.6) is $N(0, \Sigma^{-1}\Gamma_2\Sigma^{-1})$ with

$$\begin{aligned}\Gamma_2 &= (1-\delta)E\left[(\varepsilon - U^T\beta)\{X - E(X|T)\}\right]^{\otimes 2} \\ &+\delta E\left[(\varepsilon - \overline{U}^T\beta)\{X - E(X|T)\}\right]^{\otimes 2} \\ &+(1-\delta)E\left([\{UU^T - (1-\delta/2)\Sigma_{uu}\}\beta]^{\otimes 2} + UU^T\varepsilon^2\right) \\ &+\delta E\left([\{\overline{UU}^T - (1-\delta/2)\Sigma_{uu}\}\beta]^{\otimes 2} + \overline{UU}^T\varepsilon^2\right). \end{aligned} \tag{4.1.7}$$

In (4.1.7), $\overline{U}$ refers to the mean of two U's. In the case that ε is independent of (X,T), the sum of the first two terms simplifies to $\{\sigma^2 + \beta^T(1-\delta/2)\Sigma_{uu}\beta\}\Sigma$.

Standard error estimates can also be derived. A consistent estimate of Σ is

$$\widehat{\Sigma}_n = \{n - \dim(X)\}^{-1}\sum_{i=1}^{n}\left\{\overline{W}_i - \widehat{g}_{w,h}(T_i)\right\}^{\otimes 2} - (1-\delta/2)\widehat{\Sigma}_{uu}.$$

Estimates of Γ_2 are also easily developed. In the case where ε is homoscedastic and normal error, the sum of first two terms can be estimated by $(\widehat{\sigma}_n^2 + (1-\delta/2)\widehat{\beta}_n^T\widehat{\Sigma}_{uu}\widehat{\beta}_n)\widehat{\Sigma}_n$. The sum of the last two terms is a deterministic function of $(\beta, \sigma^2, \Sigma_{uu})$, and these estimates can be simply substituted into the formula.

A general sandwich-type estimator is developed as follows.

$$R_i = \widetilde{\overline{W}}_i(\widetilde{Y}_i - \widetilde{\overline{W}}_i^T\widehat{\beta}_n) + \widehat{\Sigma}_{uu}\widehat{\beta}_n/m_i + \frac{\kappa}{\delta(m_i-1)}\left\{\frac{1}{2}(W_{i1} - W_{i2})^{\otimes 2} - \widehat{\Sigma}_{uu}\right\},$$

where $\kappa = n^{-1}\sum_{i=1}^{n} m_i^{-1}$. Then a consistent estimate of Γ_2 is the sample covariance matrix of the R_i's.

4.1.5 Numerical Example

To illustrate the method, we consider data from the Framingham Heart Study. We considered $n = 1615$ males with Y being their average blood pressure in a fixed 2–year period, T being their age and W being the logarithm of the observed cholesterol level, for which there are two replicates.

We did two analyses. In the first, we used both cholesterol measurements, so that in the notation of Subsection 4.1.4, $\delta = 1$. In this analysis, there is not a great deal of measurement error. Thus, in our second analysis, which is given for illustrative purposes, we used only the first cholesterol measurement, but fixed the measurement error variance at the value obtained in the first analysis,

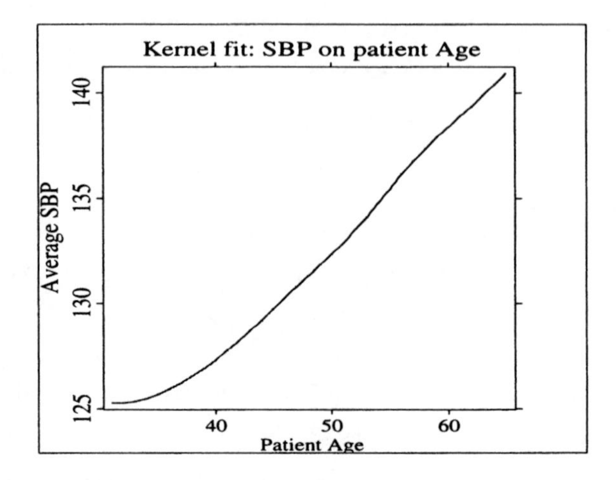

FIGURE 4.1. Estimate of the function $g(T)$ in the Framingham data ignoring measurement error.

in which $\delta = 0$. For nonparametric fitting, we chose the **bandwidth** using cross-validation to predict the response. Precisely we computed the squared error using a geometric sequence of 191 **bandwidths** ranging in $[1, 20]$. The optimal bandwidth was selected to minimize the square error among these 191 candidates. An analysis ignoring `measurement error` found some curvature in T, see Figure 4.1 for the estimate of $g(T)$.

As mentioned below, we will consider four cases and use **XploRe** (Härdle, Klinke and Müller, 1999) to calculate each case. Our results are as follows. We first consider the case in which the **measurement error** is estimated, and both `cholesterol` values are used to estimate Σ_{uu}. The estimator of β, ignoring **measurement error** was 9.438, with an estimated standard error 0.187. When we accounted for `measurement error`, the estimate increased to $\widehat{\beta} = 12.540$, and the standard error increased to 0.195.

In the second analysis, we fixed the **measurement error** variance and used only the first `cholesterol` value. The estimator of β, ignoring **measurement error**, was 10.744, with an estimated standard error 0.492. When we accounted for `measurement error`, the estimate increased to $\widehat{\beta} = 13.690$, and the standard error increased to 0.495.

4.1.6 Discussions

Our results have been phrased as if the X's were fixed constants. If they are random variables, the proofs can be simplified and the same results are obtained, now with $u_i = X_i - E(X_i|T_i)$.

The nonparametric regression estimator (4.1.4) is based on locally weighted averages. In the random X context, the same results apply if (4.1.4) is replaced by a locally linear kernel regression estimator.

If we ignore measurement error, the estimator of β is given by (1.2.2) but with the unobserved X replaced by the observed W. This differs from the correction for the attenuation estimator (4.1.2) by a simple factor which is the inverse of the reliability matrix (Gleser, 1992). In other words, the estimator which ignores measurement error is multiplied by the inverse of the reliability matrix to produce a consistent estimator of β. This same algorithm is widely employed in parametric measurement error problems for generalized linear models, where it is often known as an example of regression calibration (see Carroll, et al., 1995, for discussion and references). The use of regression calibration in our semiparametric context thus appears to be promising when (1.1.1) is replaced by a semiparametric generalized linear model.

We have treated the case in which the parametric part X of the model has measurement error and the nonparametric part T is measured exactly. An interesting problem is to interchange the roles of X and T, so that the parametric part is measured exactly and the nonparametric part is measured with error, i.e., $E(Y|X,T) = \theta T + g(X)$. Fan and Truong (1993) have discussed the case where the measurement error is normally distributed, and shown that the nonparametric function $g(\cdot)$ can be estimated only at logarithmic rates, but not with rate $n^{-2/5}$. The next section will study this problem in detail.

4.1.7 Technical Details

We first point out two facts, which will be used in the proofs.

Lemma 4.1.1 *Assume that Assumption 1.3.3 holds and that* $E(|\varepsilon_1|^4 + \|U\|^4) < \infty$. *Then*

$$n^{-1}\sum_{i=1}^{n}\varepsilon_i\left\{\sum_{j=1}^{n}\omega_{nj}(T_i)\varepsilon_j\right\} = o_P(n^{-1/2}),$$

$$n^{-1}\sum_{i=1}^{n} U_{is}\left\{\sum_{j=1}^{n}\omega_{nj}(T_i)U_{jm}\right\} = o_P(n^{-1/2}), \tag{4.1.8}$$

for $1 \le s, m \le p$.

Its proof can be completed in the same way as Lemma 5.2.3 (5.2.7). We refer the details to Liang, Härdle and Carroll (1999).

Lemma 4.1.2 *Assume that Assumptions 1.3.1-1.3.3 hold and that* $E(\varepsilon^4 + \|U\|^4) < \infty$. *Then the following equations*

$$\lim_{n\to\infty} n^{-1}\widetilde{\mathbf{W}}^T\widetilde{\mathbf{W}} = \Sigma + \Sigma_{uu} \tag{4.1.9}$$

$$\lim_{n\to\infty} n^{-1}\widetilde{\mathbf{W}}^T\widetilde{\mathbf{Y}} = \Sigma\beta \tag{4.1.10}$$

$$\lim_{n\to\infty} n^{-1}\widetilde{\mathbf{Y}}^T\widetilde{\mathbf{Y}} = \beta^T\Sigma\beta + \sigma^2 \tag{4.1.11}$$

hold in probability.

Proof. Since $W_i = X_i + U_i$ and $\widetilde{W}_i = \widetilde{X}_i + \widetilde{U}_i$, we have

$$(\widetilde{\mathbf{W}}^T\widetilde{\mathbf{W}})_{sm} = (\widetilde{\mathbf{X}}^T\widetilde{\mathbf{X}})_{sm} + (\widetilde{\mathbf{U}}^T\widetilde{\mathbf{X}})_{sm} + (\widetilde{\mathbf{X}}^T\widetilde{\mathbf{U}})_{sm} + (\widetilde{\mathbf{U}}^T\widetilde{\mathbf{U}})_{sm}. \tag{4.1.12}$$

It follows from the strong law of large numbers and Lemma A.2 that

$$n^{-1}\sum_{j=1}^{n} X_{js}U_{jm} \to 0 \quad a.s. \tag{4.1.13}$$

Observe that

$$\begin{aligned} n^{-1}\sum_{j=1}^{n}\widetilde{X}_{js}\widetilde{U}_{jm} &= n^{-1}\Big[\sum_{j=1}^{n} X_{js}U_{jm} - \sum_{j=1}^{n}\Big\{\sum_{k=1}^{n}\omega_{nk}(T_j)X_{ks}\Big\}U_{jm} \\ &\quad - \sum_{j=1}^{n}\Big\{\sum_{k=1}^{n}\omega_{nk}(T_j)U_{km}\Big\}X_{js} \\ &\quad + \sum_{j=1}^{n}\Big\{\sum_{k=1}^{n}\omega_{nk}(T_j)X_{ks}\Big\}\Big\{\sum_{k=1}^{n}\omega_{nk}(T_j)U_{km}\Big\}\Big]. \end{aligned}$$

Similar to the proof of Lemma A.3, we can prove that $\sup_{j\le n}|\sum_{k=1}^{n}\omega_{nk}(T_j)U_{km}| = o_P(1)$, which together with (4.1.13) and Assumptions 1.3.3 (ii) deduce that the above each term tends to zero. For the same reason, $n^{-1}(\widetilde{\mathbf{U}}^T\widetilde{\mathbf{X}})_{sm}$ also converges to zero.

We now prove

$$n^{-1}(\widetilde{\mathbf{U}}^T\widetilde{\mathbf{U}})_{sm} \to \sigma_{sm}^2, \tag{4.1.14}$$

where σ_{sm}^2 is the $(s,m)-$th element of Σ_{uu}. Obviously

$$\begin{aligned} n^{-1}(\widetilde{\mathbf{U}}^T\widetilde{\mathbf{U}})_{sm} = \frac{1}{n}\Big[\sum_{j=1}^n U_{js}U_{jm} & - \sum_{j=1}^n\Big\{\sum_{k=1}^n \omega_{nk}(T_j)U_{ks}\Big\}U_{jm} \\ & - \sum_{j=1}^n\Big\{\sum_{k=1}^n \omega_{nk}(T_j)U_{km}\Big\}U_{js} \\ & + \sum_{j=1}^n\Big\{\sum_{k=1}^n \omega_{nk}(T_j)U_{ks}\Big\}\Big\{\sum_{k=1}^n \omega_{nk}(T_j)U_{km}\Big\}\Big]. \end{aligned}$$

Noting that $n^{-1}\sum_{j=1}^n U_{js}U_{jm} \to \sigma_{sm}^2$, equation (4.1.14) follows from Lemma A.3 and (4.1.8). Combining (4.1.12), (4.1.14) with the arguments for $1/n(\widetilde{\mathbf{U}}^T\widetilde{\mathbf{X}})_{sm} \to 0$ and $1/n(\widetilde{\mathbf{X}}^T\widetilde{\mathbf{U}})_{sm} \to 0$, we complete the proof of (4.1.9).

Next we prove (4.1.10). It is easy to see that

$$\Big|\sum_{j=1}^n X_{js}\widetilde{g}_j\Big| \le \left(\sum_{j=1}^n X_{js}^2 \sum_{j=1}^n \widetilde{g}_j^2\right)^{1/2} \le c_n n^{1/2}\left(\sum_{j=1}^n X_{js}^2\right)^{1/2} \le Cnc_n,$$

$$\frac{1}{n}(\widetilde{\mathbf{W}}^T\widetilde{\varepsilon})_s \to 0 \text{ and } \frac{1}{n}\sum_{j=1}^n \widetilde{U}_{js}\widetilde{g}_j \to 0$$

as $n \to \infty$. Thus,

$$\begin{aligned} \frac{1}{n}(\widetilde{\mathbf{W}}^T\widetilde{\mathbf{G}})_s &= \frac{1}{n}\sum_{j=1}^n \widetilde{X}_{js}\widetilde{g}_j + \frac{1}{n}\sum_{j=1}^n \widetilde{U}_{js}\widetilde{g}_j \\ &= \frac{1}{n}\sum_{j=1}^n\Big\{X_{js} - \sum_{k=1}^n \omega_{nk}(T_j)X_{ks}\Big\}\widetilde{g}_j + \frac{1}{n}\sum_{j=1}^n \widetilde{U}_{js}\widetilde{g}_j \to 0 \end{aligned}$$

as $n \to \infty$. Combining the above arguments with (4.1.9), we complete the proof (4.1.10). The proof of (4.1.11) can be completed by the similar arguments. The details are omitted.

Proof of Theorem 4.1.1. Denote $\Delta_n = (\widetilde{\mathbf{W}}^T\widetilde{\mathbf{W}} - n\Sigma_{uu})/n$. By Lemma 4.1.2 and direct calculations,

$$\begin{aligned} n^{1/2}(\widehat{\beta}_n - \beta) &= n^{-1/2}\Delta_n^{-1}(\widetilde{\mathbf{W}}^T\widetilde{\mathbf{Y}} - \widetilde{\mathbf{W}}^T\widetilde{\mathbf{W}}\beta + n\Sigma_{uu}\beta) \\ &= n^{-1/2}\Delta_n^{-1}(\widetilde{\mathbf{X}}^T\widetilde{\mathbf{G}} + \widetilde{\mathbf{X}}^T\widetilde{\varepsilon} + \widetilde{\mathbf{U}}^T\widetilde{\mathbf{G}} + \widetilde{\mathbf{U}}^T\widetilde{\varepsilon} \\ &\qquad -\widetilde{\mathbf{X}}^T\widetilde{\mathbf{U}}\beta - \widetilde{\mathbf{U}}^T\widetilde{\mathbf{U}}\beta + n\Sigma_{uu}\beta). \end{aligned}$$

By Lemmas A.2, A.3 and 4.1.1, we conclude that

$$\begin{aligned} n^{1/2}(\widehat{\beta}_n - \beta) &= n^{-1/2}\Delta_n^{-1}\sum_{i=1}^n\Big(u_i\varepsilon_i - u_iU_i^T\beta + U_i\varepsilon_i - U_iU_i^T\beta + \Sigma_{uu}\beta\Big) + o_P(1) \\ &\stackrel{\text{def}}{=} n^{-1/2}\sum_{i=1}^n \zeta_{in} + o_P(1). \end{aligned} \quad (4.1.15)$$

Since

$$\lim_{n\to\infty} n^{-1}\sum_{i=1}^{n} u_i = 0, \ \lim_{n\to\infty} n^{-1}\sum_{i=1}^{n} u_i u_i^T = \Sigma \tag{4.1.16}$$

and $E(\varepsilon^4 + \|U\|^4) < \infty$, it follows that the k−th element $\{\zeta_{in}^{(k)}\}$ of $\{\zeta_{in}\}$ ($k = 1, \ldots, p$) satisfies that for any given $\zeta > 0$,

$$\frac{1}{n}\sum_{i=1}^{n} E\{\zeta_{in}^{(k)2} I(|\zeta_{in}^{(k)}| > \zeta n^{1/2})\} \to 0 \text{ as } n \to \infty.$$

This means that `Lindeberg's condition` for the central limit theorem holds. Moreover,

$$\begin{aligned} Cov(\zeta_{ni}) &= E\Big\{u_i(\varepsilon_i - U_i^T\beta)^2 u_i^T\Big\} + E\Big\{(U_iU_i^T - \Sigma_{uu})\beta\Big\}^{\otimes 2} + E(U_iU_i^T\varepsilon_i^2) \\ &\quad + u_i E(U_i^T\beta\beta^T U_i U_i^T) + E(U_iU_i^T\beta\beta^T U_i)u_i, \end{aligned}$$

which and (4.1.16) imply that

$$\lim_{n\to\infty} n^{-1}\sum_{i=1}^{n} Cov(\zeta_{ni}) = E(\varepsilon - U^T\beta)^2\Sigma + E\{(U\cdot U^T - \Sigma_{uu})\beta\}^{\otimes 2} + E(UU^T\varepsilon^2).$$

Theorem 4.1.1 now follows.

Proof of Theorem 4.1.2. Denote

$$A_n = n^{-1}\begin{bmatrix} \widetilde{\mathbf{Y}}^T\widetilde{\mathbf{Y}} & \widetilde{\mathbf{Y}}^T\widetilde{\mathbf{W}} \\ \widetilde{\mathbf{W}}^T\widetilde{\mathbf{Y}} & \widetilde{\mathbf{W}}^T\widetilde{\mathbf{W}} \end{bmatrix}, \qquad A = \begin{bmatrix} \beta^T\Sigma\beta + \sigma^2 & \beta^T\Sigma \\ \Sigma\beta & \Sigma + \Sigma_{uu} \end{bmatrix},$$

$$\widetilde{A}_n = n^{-1}\begin{bmatrix} (\varepsilon + U\beta)^T(\varepsilon + \mathbf{U}\beta) & (\varepsilon + U\beta)^T(U + u) \\ (U + u)^T(\varepsilon + U\beta) & (U + u)^T(U + u) \end{bmatrix}.$$

According to the definition of $\widehat{\sigma}_n^2$, a direct calculation and Lemma 4.1.1 yield that

$$\begin{aligned} n^{1/2}(\widehat{\sigma}_n^2 - \sigma^2) = n^{1/2}\sum_{j=1}^{5} S_{jn} &+ \frac{1}{n^{1/2}}(\varepsilon - \widetilde{\mathbf{U}}\beta)^T(\varepsilon - \widetilde{\mathbf{U}}\beta) \\ &- n^{1/2}(\widehat{\beta}_n^T\Sigma_{uu}\widehat{\beta}_n + \sigma^2) + o_P(1), \end{aligned}$$

where $S_{1n} = (1, -\widehat{\beta}_n^T)(A_n - \widetilde{A}_n)(1, -\widehat{\beta}_n^T)^T$, $S_{2n} = (1, -\widehat{\beta}_n^T)(\widetilde{A}_n - A)(0, \beta^T - \widehat{\beta}_n^T)^T$, $S_{3n} = (0, \beta^T - \widehat{\beta}_n^T)A(0, \beta^T - \widehat{\beta}_n^T)^T$, $S_{4n} = (0, \beta^T - \widehat{\beta}_n^T)(\widetilde{A}_n - A)(1, -\beta^T)^T$ and $S_{5n} = -(\beta - \widehat{\beta}_n)^T(\beta - \widehat{\beta}_n)$. It follows from Theorem 4.1.1 and Lemma 4.1.2 that $n^{1/2}\sum_{j=1}^{5} S_{jn} \to 0$ in probability and

$$n^{1/2}(\widehat{\sigma}_n^2 - \sigma^2) = n^{-1/2}\sum_{i=1}^{n}\Big\{(\varepsilon_i - U_i^T\beta)^2 - (\beta^T\Sigma_{uu}\beta + \sigma^2)\Big\} + o_P(1).$$

Theorem 4.1.2 now follows immediately.

Proof of Theorem 4.1.3. Since $\widehat{\beta}_n$ is a consistent estimator of β, its asymptotic bias and variance equal the relative ones of $\sum_{j=1}^{n} \omega_{nj}(t)(Y_j - W_j^T\beta)$, which is denoted by $\widehat{g}_n^*(t)$. By simple calculations,

$$\begin{aligned} E\widehat{g}_n^*(t) - g(t) &= \sum_{i=1}^{n} \omega_{ni}(t)g(T_i) - g(t), \\ E\{\widehat{g}_n^*(t) - E\widehat{g}_n^*(t)\}^2 &= \sum_{i=1}^{n} \omega_{ni}^2(t)(\beta^T \Sigma_{uu}\beta + \sigma^2). \end{aligned}$$

Theorem 4.1.3 is immediately proved.

4.2 Nonlinear Variables with Measurement Errors

4.2.1 Introduction

The previous section concerns the case where X is measured with error and T is measured exactly. In this section, we interchange the roles of X and T so that the parametric part is measured exactly and the nonparametric part is measured with error, i.e., $E(Y|X,T) = X^T\beta + g(T)$ and $W = T + U$, where U is a measurement error.

The following theorem shows that in the case of a large sample, there is no cost due to the measurement error of T when the measurement error is ordinary smooth or super smooth and X and T are independent. That is, the estimator of β given in (4.2.4), under our assumptions, is equivalent to the estimator given by (4.2.2) below when we suppose that T_i are known. This phenomenon looks unsurprising since the related work on the partially linear model suggests that a rate of at least $n^{-1/4}$ is generally needed, and since Fan and Truong (1993) proved that the nonparametric function estimate can reach the rate of $O_P(n^{-k/(2k+2\alpha+1)})$ $\{< o(n^{-1/4})\}$ in the case of ordinary smooth error. In the case of ordinary smooth error, the proof of the theorem indicates that $g(T)$ seldomly affects our estimate $\widehat{\beta}_n^*$ for the case where X and T are independent.

Fan and Truong (1993) have treated the case where $\beta = 0$ and T is observed with measurement error. They proposed a new class of kernel estimators using deconvolution and found that optimal local and global rates of convergence of these estimators depend heavily on the tail behavior of the characteristic function of the error distribution; the smoother, the slower.

In model (1.1.1), we assume that ε_i are i.i.d. and that the covariates T_i are measured with errors, and we can only observe their surrogates W_i, i.e.,

$$W_i = T_i + U_i, \tag{4.2.1}$$

where the `measurement errors` U_i are i.i.d., independent of (Y_i, X_i, T_i), with mean zero and covariance matrix Σ_{uu}. We will assume that U has a known distribution, which was proposed by Fan and Truong (1993) to assure that the model is identifiable. The model (1.1.1) with (4.2.1) can be seen as a mixture of linear and nonlinear `errors-in-variables` models.

Recalling the argument for (1.2.2), it can be obtained as follows. Let $\widehat{g}_{y,h}(\cdot)$ and $\widehat{g}_{x,h}(\cdot)$ be the kernel regression estimators of $E(Y|T)$ and $E(X|T)$, respectively. Then

$$\beta_{LS} = \Big[\sum_{i=1}^{n}\{X_i - \widehat{g}_{x,h}(T_i)\}^{\otimes 2}\Big]^{-1}\sum_{i=1}^{n}\{X_i - \widehat{g}_{x,h}(T_i)\}\{Y_i - \widehat{g}_{y,h}(T_i)\}. \tag{4.2.2}$$

Due to the disturbance of `measurement error` U and the fact that $\widehat{g}_{x,h}(T_i)$ and $\widehat{g}_{y,h}(T_i)$ are no longer statistics, the least squares form of (4.2.2) must be modified. In the next subsection, we will redefine an estimator of β. More exactly, we have to find a new estimator of $g(\cdot)$ and then perform the regression of Y and X on W. The asymptotic normality of the resulting estimator of β depends on the smoothness of the error distribution.

4.2.2 Construction of Estimators

As pointed out in the former subsection, our first objective is to estimate the nonparametric function $g(\cdot)$ when T is observed with error. This can be overcome by using the ideas of Fan and Truong (1993). By using the `deconvolution` technique, one can construct consistent nonparametric estimates of $g(\cdot)$ with some convergence rate under appropriate assumptions. First, we briefly describe the `deconvolution` method, which has been studied by Stefanski and Carroll (1990), and Fan and Truong (1993). Denote the densities of W and T by $f_W(\cdot)$ and $f_T(\cdot)$, respectively. As pointed out in the literature, $f_T(\cdot)$ can be estimated by

$$\widehat{f}_n(t) = \frac{1}{nh_n}\sum_{j=1}^{n} K_n\Big(\frac{t - W_j}{h_n}\Big)$$

with

$$K_n(t) = \frac{1}{2\pi}\int_{R^1} \exp(-ist)\frac{\phi_K(s)}{\phi_U(s/h_n)}ds, \tag{4.2.3}$$

where $\phi_K(\cdot)$ is the Fourier transform of $K(\cdot)$, a kernel function and $\phi_U(\cdot)$ is the characteristic function of the error variable U. For a detailed discussion, see Fan and Truong (1993). Denote

$$\omega_{ni}^*(\cdot) = K_n\Big(\frac{\cdot - W_i}{h_n}\Big)\Big/\sum_j K_n\Big(\frac{\cdot - W_j}{h_n}\Big) \stackrel{\text{def}}{=} \frac{1}{nh_n}K_n\Big(\frac{\cdot - W_i}{h_n}\Big)\Big/\widehat{f}_n(\cdot).$$

Now let us return to our goal. Replacing the $g_n(t)$ in Section 1.2 by

$$g_n^*(t) = \sum_{i=1}^n \omega_{ni}^*(t)(Y_i - X_i^T\beta),$$

then the least squares estimator $\widehat{\beta}_n^*$ of β can be explicitly expressed as

$$\widehat{\beta}_n^* = (\widetilde{\mathbf{X}}^T\widetilde{\mathbf{X}})^{-1}(\widetilde{\mathbf{X}}^T\widetilde{\mathbf{Y}}), \tag{4.2.4}$$

where $\widetilde{\mathbf{Y}}$ denotes $(\widetilde{Y}_1, \ldots, \widetilde{Y}_n)$ with $\widetilde{Y}_i = Y_i - \sum_{j=1}^n \omega_{nj}^*(W_i)Y_j$ and $\widetilde{\mathbf{X}}$ denotes $(\widetilde{X}_1, \ldots, \widetilde{X}_n)$ with $\widetilde{X}_i = X_i - \sum_{j=1}^n \omega_{nj}^*(W_i)X_j$. The estimator $\widehat{\beta}_n^*$ will be shown to possess asymptotic normality under appropriate conditions.

4.2.3 Asymptotic Normality

Assumption 4.2.1 *(i) The marginal density $f_T(\cdot)$ of the unobserved covariate T is bounded away from 0 on $[0,1]$, and has a bounded $k-$th derivative, where k is a positive integer. (ii) The characteristic function of the error distribution $\phi_U(\cdot)$ does not vanish. (iii) The distribution of the error U is* `ordinary smooth` *or* `super smooth`.

The definitions of `super smooth` and `ordinary smooth` distributions were given by Fan and Truong (1993). We also state them here for easy reference.

1. `Super smooth` of order α: *If the characteristic function of the error distribution $\phi_U(\cdot)$ satisfies*

$$d_0|t|^{\alpha_0}\exp(-|t|^\alpha/\zeta) \le |\phi_U(t)| \le d_1|t|^{\alpha_1}\exp(-|t|^\alpha/\zeta) \ \ as\ t \to \infty, \tag{4.2.5}$$

where d_0, d_1, α and ζ are positive constants, and α_0 and α_1 are constants.

2. `Ordinary smooth` of order α: *If the characteristic function of the error distribution $\phi_U(\cdot)$ satisfies*

$$d_0|t|^{-\alpha} \le |\phi_U(t)| \le d_1|t|^{-\alpha} \quad as\ t \to \infty, \tag{4.2.6}$$

for positive constants d_0, d_1 and α.

For example, standard normal and Cauchy distributions are **super smooth** with $\alpha = 2$ and $\alpha = 1$ respectively. The gamma distribution of degree p and the double exponential distribution are **ordinary smooth** with $\alpha = p$ and $\alpha = 2$, respectively. We should note that an error cannot be both ordinary smooth and super smooth.

Assumption 4.2.2 *(i) The regression functions $g(\cdot)$ and $h_j(\cdot)$ have continuous kth derivatives on $[0,1]$.*

(ii) The kernel $K(\cdot)$ is a $k-th$ order kernel function, that is

$$\int_{-\infty}^{\infty} K(u)du = 1, \quad \int_{-\infty}^{\infty} u^l K(u)du \begin{cases} = 0 & l = 1, \dots, k-1, \\ \neq 0 & l = k. \end{cases}$$

Assumption 4.2.2 (i) modifies Assumption 1.3.2 to meet the condition 1 of Fan and Truong (1993).

Our main result is concerned with the limit distribution of the estimate of β stated as follows.

Theorem 4.2.1 *Suppose that Assumptions 1.3.1, 4.2.1 and 4.2.2 hold and that $E(|\varepsilon|^3 + \|U\|^3) < \infty$. If either of the following conditions holds, then $\widehat{\beta}_n^*$ is an asymptotically normal estimator, i.e., $n^{1/2}(\widehat{\beta}_n - \beta) \longrightarrow^{\mathcal{L}} N(0, \sigma^2\Sigma^{-1})$.*

(i) The error distribution is **super smooth**. *X and T are mutually independent. $\phi_K(t)$ has a bounded support on $|t| \le M_0$. We take the* **bandwidth** *$h_n = c(\log n)^{-1/\alpha}$ with $c > M_0(2/\zeta)^{1/\alpha}$;*

(ii) The error distribution is **ordinary smooth**. *We take $h_n = dn^{-1/(2k+2\alpha+1)}$ with $d > 0$ and $2k > 2\alpha + 1$.*

$$t^\alpha \phi_U(t) \to c, \quad t^{\alpha+1}\phi_U'(t) = O(1) \quad \text{as } t \to \infty$$

for some constant $c \neq 0$,

$$\int_{-\infty}^{\infty} |t|^{\alpha+1}\{\phi_K(t) + \phi_K'(t)\}dt < \infty, \qquad \int_{-\infty}^{\infty} |t^{\alpha+1}\phi_K(t)|^2 dt < \infty.$$

4.2.4 Simulation Investigations

We conduct a moderate sample Monte-Carlo simulation to show the behavior of the estimator $\widehat{\beta}_n^*$. A generalization of the model studied by Fan and Truong (1993) is considered.

$$Y = X^T\beta + g(T) + \varepsilon \text{ and } W = T + U \quad \text{with } \beta = 0.75,$$

where $X \sim N(0,1)$, $T \sim N(0.5, 0.25^2)$, $\varepsilon \sim N(0, 0.0015^2)$ and $g(t) = t_+^3(1-t)_+^3$. Two kinds of error distributions are examined to study the effect of them on the mean squared error (MSE) of the estimator $\widehat{\beta}_n^*$: one is normal and the other is double exponential.

1. (Double exponential error). U has a double exponential distribution:

$$f_U(u) = (\sqrt{2}\sigma_0)^{-1}\exp(-\sqrt{2}|u|/\sigma_0) \quad \text{for } \sigma_0^2 = (3/7)\mathrm{Var}(T).$$

   Let $K(\cdot)$ be the Gaussian kernel

$$K(x) = (\sqrt{2\pi})^{-1}\exp(-x^2/2),$$

   then

$$K_n(x) = (\sqrt{2\pi})^{-1}\exp(-x^2/2)\Big\{1 - \frac{\sigma_0^2}{2h_n^2}(x^2-1)\Big\}. \tag{4.2.7}$$

2. (Normal error). $U \sim N(0, 0.125^2)$. Suppose the function $K(\cdot)$ has a Fourier transform by $\phi_K(t) = (1-t^2)_+^2$. By (4.2.3),

$$K_n(t) = \frac{1}{\pi}\int_0^1 \cos(st)(1-s^2)^3\exp\Big(\frac{0.125^2 s^2}{2h_n^2}\Big)ds. \tag{4.2.8}$$

For the above model, we use three different kernels: (4.2.7), (4.2.8) and `quartic kernel` $(15/16)(1-u^2)^2 I(|u| \le 1)$ (ignoring `measurement error`). Our aim is to compare the results in the cases of considering measurement error and ignoring `measurement error`. The results for different sample numbers are presented in $N = 2000$ replications. The mean square errors (MSE) are calculated based on 100, 500, 1000 and 2000 observations with three kinds of kernels. Table 4.1 gives the final detailed simulation results. The simulations reported show that the behavior of MSE with double exponential error model is the best one, while the behavior of MSE with quartic kernel is the worst one.

In the simulation procedure, we also fit the nonparametric part using

$$\widehat{g}_n^*(t) = \sum_{i=1}^n \omega_{ni}^*(t)(Y_i - X_i^T\widehat{\beta}_n^*), \tag{4.2.9}$$

where $\widehat{\beta}_n^*$ is the resulting estimator given in (4.2.4).

An analysis ignoring `measurement error` (with `quartic kernel`) finds some curvature in T. See Figure 4.2 for the comparison of $g(T)$ with its estimator

TABLE 4.1. MSE($\times 10^{-3}$) of the estimator $\widehat{\beta}_n^*$

Kernel	$n = 100$	$n = 500$	$n = 1000$	$n = 2000$
	MSE	MSE	MSE	MSE
(4.2.7)	3.095000	0.578765	0.280809	0.151008
(4.2.8)	7.950000	1.486650	0.721303	0.387888
quartic	15.52743	10.36125	8.274210	4.166037

(4.2.9) using the different-size samples. Each curve represents the mean of 2000 realizations of these true curves and estimating curves. The solid lines stand for true values and the dashed lines stand for the values of the resulting estimator given by (4.2.9).

The bandwidth used in our simulation is selected using cross-validation to predict the response. More precisely, we compute the average squared error using a geometric sequence of 41 bandwidths ranging in $[0.1, 0.5]$. The optimal bandwidth is selected to minimize the average squared error among 41 candidates. The results reported here support our theoretical procedure, and illustrate that our estimators for both the parametric and nonparametric parts work very well numerically.

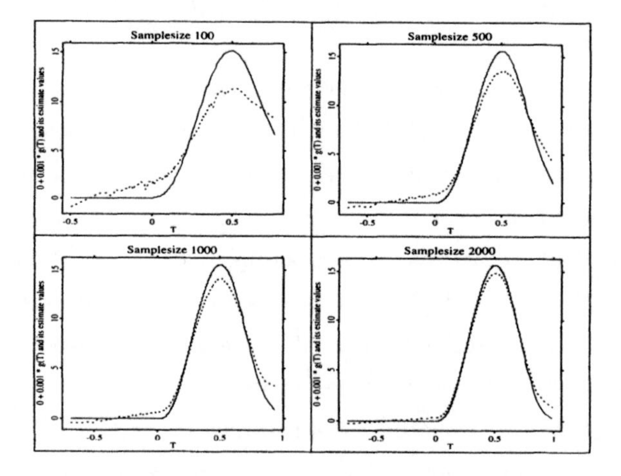

FIGURE 4.2. Estimates of the function $g(T)$.

4.2.5 Technical Details

Lemma 4.2.1 provides bounds for $h_j(T_i) - \sum_{k=1}^n \omega_{nk}^*(W_i) h_j(T_k)$ and $g(T_i) - \sum_{k=1}^n \omega_{nk}^*(W_i) g(T_k)$. The proof is partly based upon the conclusion of Fan and Truong

(1993).

Lemma 4.2.1 *Suppose that Assumptions 1.3.1 and 4.2.2 hold. Then for all* $1 \leq l \leq p$ *and* $1 \leq i, j \leq p$

$$E(\widetilde{g}_i^* \widetilde{g}_j^* \widetilde{h}_{il}^* \widetilde{h}_{jl}^*) = O(h^{4k}),$$

where

$$\begin{aligned} \widetilde{g}_i^* &= \frac{1}{f_T(W_i)} \sum_{k=1}^{n} \{g(T_i) - g(T_k)\} \frac{1}{nh_n} K_n\Big(\frac{W_i - W_k}{h_n}\Big), \\ \widetilde{h}_{il}^* &= \frac{1}{f_T(W_i)} \sum_{k=1}^{n} \{h_l(T_i) - h_l(T_k)\} \frac{1}{nh_n} K_n\Big(\frac{W_i - W_k}{h_n}\Big). \end{aligned}$$

Proof. Similar to the proof of Lemma 1 of Fan and Truong (1993), we have

$$\begin{aligned} E(\widetilde{g}_i^* \widetilde{g}_j^* \widetilde{h}_{il}^* \widetilde{h}_{jl}^*) &= \frac{1}{h_n^4} \underbrace{\int_{-\infty}^{\infty} \cdots \int_{-\infty}^{\infty}}_{8} \{g(u_1) - g(u_2)\}\{g(u_3) - g(u_4)\}\{h_l(v_1) - h_l(v_2)\} \\ \{h_l(v_3) &- h_l(v_4)\} \times K_n\Big(\frac{u_1 - u_2}{h_n}\Big) K_n\Big(\frac{u_3 - u_4}{h_n}\Big) K_n\Big(\frac{v_1 - v_2}{h_n}\Big) K_n\Big(\frac{v_3 - v_4}{h_n}\Big) \\ & f_T(u_2) f_T(u_4) f_T(v_2) f_T(v_4) \prod_{q=1}^{4} du_q \, dv_q \\ &= O(h_n^{4k}) \end{aligned}$$

by applying Assumption 4.2.2.

Proof of Theorem 4.2.1. We first outline the proof of the theorem. We decompose $\sqrt{n}(\beta_n - \beta)$ into three terms. Then we calculate the tail probability value of each term. By the definition of $\widehat{\beta}_n^*$,

$$\begin{aligned} \sqrt{n}(\widehat{\beta}_n^* - \beta) &= \sqrt{n}(\widetilde{\mathbf{X}}^T \widetilde{\mathbf{X}})^{-1} \Big[\sum_{i=1}^{n} \widetilde{X}_i \widetilde{g}_i - \sum_{i=1}^{n} \widetilde{X}_i \Big\{\sum_{j=1}^{n} \omega_{nj}^*(W_i) \varepsilon_j\Big\} + \sum_{i=1}^{n} \widetilde{X}_i \varepsilon_i\Big] \\ &\stackrel{\text{def}}{=} A(n) \Big[\frac{1}{\sqrt{n}} \sum_{i=1}^{n} \widetilde{X}_i \widetilde{g}_i - \frac{1}{\sqrt{n}} \sum_{i=1}^{n} \widetilde{X}_i \Big\{\sum_{j=1}^{n} \omega_{nj}^*(W_i) \varepsilon_j\Big\} \\ &\quad + \frac{1}{\sqrt{n}} \sum_{i=1}^{n} \widetilde{X}_i \varepsilon_i\Big], \end{aligned} \tag{4.2.10}$$

where $A(n) = n^{-1} \widetilde{\mathbf{X}}^T \widetilde{\mathbf{X}}$ and $\widetilde{g}_i = g(T_i) - \sum_{k=1}^{n} \omega_{nk}^*(W_i) g(T_k)$. Similar to Lemma A.2 below, we can show that $A(n)$ converges to Σ^{-1} in probability.

In view of (4.2.10), in order to prove Theorem 4.2.1, it suffices to show that

$$\sum_{k=1}^{n} \widetilde{X}_i \widetilde{g}_i = o_P(\sqrt{n}), \tag{4.2.11}$$

$$\sum_{i=1}^{n} \widetilde{X}_i \Big\{\sum_{j=1}^{n} \omega_{nj}^*(W_i) \varepsilon_j\Big\} = o_P(\sqrt{n}), \tag{4.2.12}$$

and

$$\frac{1}{\sqrt{n}}\sum_{i=1}^{n}\widetilde{X}_i\varepsilon_i \longrightarrow^{\mathcal{L}} N(0,\sigma^2\Sigma). \tag{4.2.13}$$

Obviously for all $1 \le j \le p$

$$\sum_{i=1}^{n}\widetilde{X}_{ij}\widetilde{g}_i = \sum_{i=1}^{n}u_{ij}\widetilde{g}_i + \sum_{i=1}^{n}\widetilde{h}_{nj}(T_i)\widetilde{g}_i - \sum_{i=1}^{n}\sum_{q=1}^{n}\omega^*_{nq}(W_i)u_{qj}\widetilde{g}_i, \tag{4.2.14}$$

where $\widetilde{h}_{nj}(T_i) = h_j(T_i) - \sum_{k=1}^{n}\omega^*_{nk}(W_i)h_j(T_k)$.

Similar to the proof of Lemma A.3, we can prove that for $1 \le j \le p$,

$$\max_{1\le i\le n}\Big|\sum_{k=1}^{n}\omega^*_{nk}(W_i)\varepsilon_k\Big| = o_P(n^{-1/4}) \tag{4.2.15}$$

$$\max_{1\le i\le n}\Big|\sum_{k=1}^{n}\omega^*_{nk}(W_i)u_{kj}\Big| = o_P(n^{-1/4}). \tag{4.2.16}$$

Equations (4.2.15) and (4.2.16) will be used repeatedly in the following proof. Taking $r = 3$, $V_k = u_{kl}$ of u_{kj}, $a_{ji} = \omega^*_{ni}(W_i)$, $p_1 = 2/3$, and $p_2 = 0$ in Lemma A.3 below, we can prove both (4.2.15) and (4.2.16).

For the case where U is ordinary smooth error. Analogous to the proof of Lemma A.3, we can prove that

$$\sum_{i=1}^{n}u_{ij}\widetilde{g}_i = o_P(n^{1/2}). \tag{4.2.17}$$

Similarly,

$$\Big|\sum_{i=1}^{n}\Big\{\sum_{q=1}^{n}\omega^*_{nq}(W_i)u_{qj}\Big\}\widetilde{g}_i\Big| \le n\max_{i\le n}|\widetilde{g}_i|\max_{i\le n}\Big|\sum_{q=1}^{n}\omega^*_{nq}(W_i)u_{qj}\Big| = o_P(n^{1/2}). \tag{4.2.18}$$

In view of (4.2.14), (4.2.17) and (4.2.18), in order to prove (4.2.11), it suffices to show

$$\sum_{i=1}^{n}\widetilde{h}_{nj}(T_i)\widetilde{g}_i = o_P(n^{1/2})$$

which is equivalent to

$$\sum_{k=1}^{n}\sum_{l=1}^{n}\sum_{i=1}^{n}\{g(T_i)-g(T_k)\}\{h_j(T_i)-h_j(T_l)\}\omega_{nk}(T_i)\omega_{nl}(T_i) = o_P(n^{1/2}). \tag{4.2.19}$$

In order to prove (4.2.19), noting that

$$\sup_t|\widehat{f}_n(t) - f_T(t)| = o_P(1)$$

which is similar to Lemma 2 of Fan and Truong (1993), it suffices to show

$$\begin{aligned}
&\sum_{i=1}^{n}\sum_{k=1}^{n}\sum_{l=1}^{n}\{g(T_i)-g(T_k)\}\{h_j(T_i)-h_j(T_l)\}\frac{1}{f_T^2(W_i)} \\
&\qquad \frac{1}{n^2h_n^2}K_n\Big(\frac{W_i-W_k}{h_n}\Big)K_n\Big(\frac{W_i-W_l}{h_n}\Big) \\
&= \sum_{i=1}^{n}\widetilde{g}_i^*\widetilde{h}_{ij}^* = o_P(\sqrt{n}),
\end{aligned}$$

which follows from for any given $\delta>0$

$$\begin{aligned}
P\Big(\Big|\sum_{i=1}^{n}\widetilde{g}_i^*\widetilde{h}_{ij}^*\Big|>\delta\sqrt{n}\Big) &\leq \frac{1}{n\delta^2}E\Big(\sum_{i=1}^{n}\widetilde{g}_i^*\widetilde{h}_{ij}^*\Big)^2 \\
&= \frac{1}{n\delta^2}\Big\{\sum_{i=1}^{n}E(\widetilde{g}_i^*\widetilde{h}_{ij}^*)^2+\sum_{i=1}^{n}\sum_{k=1,k\neq i}^{n}E(\widetilde{g}_i^*\widetilde{g}_k^*\widetilde{h}_{ij}^*\widetilde{h}_{kj}^*)\Big\} \\
&= \frac{1}{n\delta^2}\{O(nh_n^{4k})+O(n^2h_n^{4k})\}=o(1). \qquad (4.2.20)
\end{aligned}$$

The last step uses Lemma 4.2.1 and the fact that $h_n=dn^{-1/(2k+2\alpha+1)}$ with $d>0$ and $2k>2\alpha+1$.

Thus, equations (4.2.17), (4.2.18) and (4.2.20) imply (4.2.11).

Observe that

$$\begin{aligned}
\sum_{i=1}^{n}\Big\{\sum_{k=1}^{n}\widetilde{X}_{kj}\omega_{ni}^*(W_k)\Big\}\varepsilon_i &= \sum_{i=1}^{n}\Big\{\sum_{k=1}^{n}u_{kj}\omega_{ni}^*(W_k)\Big\}\varepsilon_i \\
&\quad +\sum_{i=1}^{n}\Big\{\sum_{k=1}^{n}\widetilde{h}_{nj}(T_k)\omega_{ni}^*(W_k)\Big\}\varepsilon_i \\
&\quad -\sum_{i=1}^{n}\Big[\sum_{k=1}^{n}\Big\{\sum_{q=1}^{n}u_{qj}\omega_{nq}^*(W_k)\Big\}\omega_{ni}^*(W_k)\Big]\varepsilon_i. (4.2.21)
\end{aligned}$$

In order to prove (4.2.12), it suffices to show

$$\sum_{i=1}^{n}\Big\{\sum_{k=1}^{n}u_{kj}\omega_{ni}^*(W_k)\Big\}\varepsilon_i=o_p(n^{1/2}) \qquad (4.2.22)$$

$$\sum_{i=1}^{n}\Big\{\sum_{k=1}^{n}\widetilde{h}_{nk}(T_j)\omega_{ni}^*(W_k)\Big\}\varepsilon_i=o_p(n^{1/2}) \qquad (4.2.23)$$

$$\sum_{i=1}^{n}\Big[\sum_{k=1}^{n}\Big\{\sum_{q=1}^{n}u_{qj}\omega_{nq}^*(W_k)\Big\}\omega_{ni}^*(W_k)\Big]\varepsilon_i=o_p(n^{1/2}) \qquad (4.2.24)$$

Applying (4.2.15) and (4.2.16), equation (4.2.24) follows from

$$\begin{aligned}
\Big|\sum_{i=1}^{n}\Big[\sum_{k=1}^{n}\Big\{\sum_{q=1}^{n}u_{qj}\omega_{nq}^*(W_k)\Big\}\omega_{ni}^*(W_k)\Big]\varepsilon_i\Big| &\leq n\max_{k\leq n}\Big|\sum_{i=1}^{n}\omega_{ni}^*(W_k)\varepsilon_i\Big| \\
&\quad \times\max_{k\leq n}\Big|\sum_{q=1}^{k}u_{qj}\omega_{nq}^*(W_j)\Big| \\
&= o_p(n^{1/2}).
\end{aligned}$$

Similar to the proof of Lemma A.3 below, we finish the proof of (4.2.22).

Analogous to (4.2.19)-(4.2.20), in order to prove (4.2.23), it suffices to show that

$$\frac{1}{nh_n}\sum_{i=1}^{n}\Big\{\sum_{k=1}^{n}\frac{1}{f_T(W_i)}K_n\Big(\frac{W_k-W_i}{h_n}\Big)\widetilde{h}_{kj}^*\Big\}\varepsilon_i=o_P(n^{1/2}),$$

which follows from

$$\frac{1}{n^2h_n^2}\sum_{i=1}^{n}E\Big\{\sum_{k=1}^{n}\frac{1}{f_T(W_i)}K_n\Big(\frac{W_k-W_i}{h_n}\Big)\widetilde{h}_{kj}^*\Big\}^2E\varepsilon_i^2=o(n).$$

Thus, the proof of (4.2.12) follows from (4.2.22)-(4.2.24.)

The proof of (4.2.13) follows from CLT and the fact that $1/n\widetilde{\mathbf{X}}^T\widetilde{\mathbf{X}}\to\Sigma$ holds in probability as $n\to\infty$.

When U is a super smooth error. By checking the above proofs, we find that the proof of (4.2.11) is required to be modified due to the fact that both (4.2.17) and (4.2.20) are no longer true when U is a super smooth error.

Similar to (4.2.19)-(4.2.20), in order to prove (4.2.11), it suffices to show that

$$\begin{aligned}&\frac{1}{n^2h_n^2}\sum_{i=1}^{n}\Big\{\sum_{k=1}^{n}(X_{ij}-X_{kj})\frac{1}{f_T(W_i)}K_n\Big(\frac{W_i-W_k}{h_n}\Big)\Big\}\\&\quad\Big\{\sum_{k=1}^{n}\{g(T_i)-g(T_k)\}\frac{1}{f_T(W_i)}K_n\Big(\frac{W_i-W_k}{h_n}\Big)\Big\}\\&\quad=o_P(\sqrt{n}).\end{aligned}$$

Let

$$\begin{aligned}\widetilde{X}_{ij}^*&=\frac{1}{nh_n}\sum_{k=1}^{n}(X_{ij}-X_{kj})\frac{1}{f_T(W_i)}K_n\Big(\frac{W_i-W_k}{h_n}\Big)\\\widetilde{g}_i^*&=\frac{1}{nh_n}\sum_{k=1}^{n}\{g(T_i)-g(T_k)\}\frac{1}{f_T(W_i)}K_n\Big(\frac{W_i-W_k}{h_n}\Big).\end{aligned}$$

Observe that

$$E\Big(\sum_{i=1}^{n}\widetilde{X}_{ij}^*\widetilde{g}_i^*\Big)^2=\sum_{i=1}^{n}E(\widetilde{X}_{ij}^*\widetilde{g}_i^*)^2+\sum_{i=1}^{n}\sum_{k=1,k\neq i}^{n}E(\widetilde{X}_{ij}^*\widetilde{g}_i^*\widetilde{X}_{kj}^*\widetilde{g}_k^*).$$

Since X_i and (T_i,W_i) are independent, we have for $j=1$, $\widetilde{\mathbf{T}}=(T_1,\cdots,T_n)$ and $\widetilde{\mathbf{W}}=(W_1,\cdots,W_n)$,

$$E(\widetilde{X}_{i1}^*\widetilde{X}_{k1}^*|\widetilde{\mathbf{T}},\widetilde{\mathbf{W}})=E\{(X_{11}-X_{21})(X_{31}-X_{21})\}\frac{1}{n^2h_n^2}\sum_{l=1,l\neq i,k}$$

$$K_n\Big(\frac{W_i - W_l}{h_n}\Big)K_n\Big(\frac{W_k - W_l}{h_n}\Big)$$
$$+E(X_{11} - X_{21})^2\frac{1}{n^2h_n^2}K_n\Big(\frac{W_i - W_k}{h_n}\Big)K_n\Big(\frac{W_k - W_i}{h_n}\Big)$$
$$+E\{(X_{11} - X_{21})(X_{31} - X_{11})\}\frac{1}{n^2h_n^2}\sum_{l=1,l\neq i,k} K_n\Big(\frac{W_i - W_l}{h_n}\Big)K_n\Big(\frac{W_k - W_i}{h_n}\Big)$$
$$+E\{(X_{11} - X_{21})(X_{21} - X_{31})\}\frac{1}{n^2h_n^2}\sum_{l=1,l\neq i,k} K_n\Big(\frac{W_i - W_k}{h_n}\Big)K_n\Big(\frac{W_k - W_l}{h_n}\Big)$$

for all $1 \le i, k \le n$.

Similar to the proof of Lemma 4.2.1, we can show that for all $1 \le j \le p$

$$E\Big(\sum_{i=1}^{n} \widetilde{X}_{ij}^{*}\widetilde{g}_i^{*}\Big)^2 = O(nh_n).$$

Finally, for any given $\eta > 0$

$$P\Big\{\Big|\sum_{i=1}^{n} \widetilde{X}_{ij}^{*}\widetilde{g}_i^{*}\Big| > \sqrt{n}\eta\Big\} \le \frac{1}{n\eta^2}E\Big(\sum_{i=1}^{n} \widetilde{X}_{ij}^{*}\widetilde{g}_i^{*}\Big)^2 = O(h_n) \to 0$$

as $n \to \infty$, which implies (4.2.23) and therefore the proof of Theorem 4.2.1 is completed.

5 SOME RELATED THEORETIC TOPICS

5.1 The Laws of the Iterated Logarithm

5.1.1 Introduction

The CLT states that the estimator given in (1.2.2) converges to normal distribution, but does not provide information about the fluctuations of this estimator about the true value. The `laws of iterative logarithm (LIL)` complement the CLT by describing the precise extremes of the fluctuations of the estimator (1.2.2). This forms the key of this section. The following studies assert that the extreme fluctuations of the estimators (1.2.2) and (1.2.4) are essentially the same order of magnitude $(2\log\log n)^{1/2}$ as classical LIL for the i.i.d. case.

The aim of this section is concerned with the case where ε_i are i.i.d. and (X_i, T_i) are `fixed design points`. Similar results for the `random design` case can be found in Hong and Cheng (1992a) and Gao (1995a, b). The version of the LIL for the estimators is stated as follows:

Theorem 5.1.1 *Suppose that Assumptions 1.3.1-1.3.3 hold. If* $E|\varepsilon_1|^3 < \infty$, *then*

$$\limsup_{n\to\infty}\left(\frac{n}{2\log\log n}\right)^{1/2}|\beta_{LSj}-\beta_j| = (\sigma^2\sigma^{jj})^{1/2}, \quad a.s. \tag{5.1.1}$$

Furthermore, let $b_n = Cn^{-3/4}(\log n)^{-1}$ *in Assumption 1.3.3 and if* $E\varepsilon_1^4 < \infty$, *then*

$$\limsup_{n\to\infty}\left(\frac{n}{2\log\log n}\right)^{1/2}|\widehat{\sigma}_n^2-\sigma^2| = (Var\varepsilon_1^2)^{1/2}, \quad a.s. \tag{5.1.2}$$

where β_{LSj}, β_j *and* σ^{jj} *denote the* $j-th$ *element of* β_{LS}, β *and the* $(j,j)-th$ *element of* Σ^{-1} *respectively.*

We outline the proof of the theorem. First we decompose $\sqrt{n}(\beta_{LS}-\beta)$ and $\sqrt{n}(\widehat{\sigma}_n^2-\sigma^2)$ into three terms and five terms respectively. Then we calculate the `tail probability` value of each term. We have, from the definitions of β_{LS} and $\widehat{\sigma}_n^2$, that

$$\begin{aligned}\sqrt{n}(\beta_{LS}-\beta) &= \sqrt{n}(\widetilde{\mathbf{X}}^T\widetilde{\mathbf{X}})^{-1}\Big\{\sum_{i=1}^n\widetilde{X}_i\widetilde{g}_i - \sum_{i=1}^n\widetilde{X}_i\sum_{j=1}^n\omega_{nj}(T_i)\varepsilon_j + \sum_{i=1}^n\widetilde{X}_i\varepsilon_i\Big\}\\ &\stackrel{\text{def}}{=} n(\widetilde{\mathbf{X}}^T\widetilde{\mathbf{X}})^{-1}(H_1-H_2+H_3);\end{aligned} \tag{5.1.3}$$

$$\begin{aligned}
\sqrt{n}(\widehat{\sigma}_n^2 - \sigma^2) &= \frac{1}{\sqrt{n}}\widetilde{\mathbf{Y}}^T\{\mathcal{F} - \widetilde{\mathbf{X}}(\widetilde{\mathbf{X}}^T\widetilde{\mathbf{X}})^{-1}\widetilde{\mathbf{X}}^T\}\widetilde{\mathbf{Y}} - \sqrt{n}\sigma^2 \\
&= \frac{1}{\sqrt{n}}\varepsilon^T\varepsilon - \sqrt{n}\sigma^2 - \frac{1}{\sqrt{n}}\varepsilon^T\widetilde{\mathbf{X}}(\widetilde{\mathbf{X}}^T\widetilde{\mathbf{X}})^{-1}\widetilde{\mathbf{X}}^T\varepsilon \\
&\quad + \frac{1}{\sqrt{n}}\widehat{\mathbf{G}}^T\{\mathcal{F} - \widetilde{\mathbf{X}}(\widetilde{\mathbf{X}}^T\widetilde{\mathbf{X}})^{-1}\widetilde{\mathbf{X}}^T\}\widehat{\mathbf{G}} \\
&\quad - \frac{2}{\sqrt{n}}\widehat{\mathbf{G}}^T\widetilde{\mathbf{X}}(\widetilde{\mathbf{X}}^T\widetilde{\mathbf{X}})^{-1}\widetilde{\mathbf{X}}^T\varepsilon + \frac{2}{\sqrt{n}}\widehat{\mathbf{G}}^T\varepsilon \\
&\stackrel{\text{def}}{=} \sqrt{n}\{(I_1 - \sigma^2) - I_2 + I_3 - 2I_4 + 2I_5\}, \qquad (5.1.4)
\end{aligned}$$

where $\widehat{\mathbf{G}} = \{g(T_1) - \widehat{g}_n(T_1), \ldots, g(T_n) - \widehat{g}_n(T_n)\}^T$ and $\widehat{g}_n(\cdot)$ is given by (1.2.3).

In the following steps we prove that each element of H_1 and H_2 converges almost surely to zero, and $\sqrt{n}I_i$ also converge almost surely to zero for $i = 2, 3, 4, 5$. The proof of the first half assertion will be finished in steps 1 and 2. The proof of the second half assertion will be arranged in Subsection 5.1.3 after we complete the proof of (5.1.1). Finally, we apply Corollary 5.2.3 of Stout (1974) to complete the proof of the theorem.

5.1.2 Preliminary Processes

Step 1.

$$\sqrt{n}H_{1j} = \sqrt{n}\sum_{i=1}^n \widetilde{x}_{ij}\widetilde{g}_i = O(n^{1/2}\log^{-1/2} n) \quad \text{for } j = 1, \ldots, p. \qquad (5.1.5)$$

Proof. Recalling the definition of h_{nij} given on the page 29, $\sqrt{n}H_{1j}$ can be decomposed into

$$\sum_{i=1}^n u_{ij}\widetilde{g}_i + \sum_{i=1}^n h_{nij}\widetilde{g}_i - \sum_{i=1}^n\sum_{q=1}^n \omega_{nq}(T_i)u_{qj}\widetilde{g}_i.$$

By Lemma A.1,

$$\Big|\sum_{i=1}^n h_{nij}\widetilde{g}_i\Big| \le n\max_{i\le n}|\widetilde{g}_i|\max_{i\le n}|h_{nij}| = O(nc_n^2).$$

Applying Abel's inequality, and using (1.3.1) and Lemma A.1 below, we have for all $1 \le j \le p$,

$$\sum_{i=1}^n u_{ij}\widetilde{g}_i = O(n^{1/2}\log nc_n)$$

and

$$\begin{aligned}
\Big|\sum_{i=1}^n\sum_{q=1}^n \omega_{nq}(T_i)u_{qj}\widetilde{g}_i\Big| &\le n\max_{i\le n}|\widetilde{g}_i|\max_{i\le n}\Big|\sum_{q=1}^n \omega_{nq}(T_i)u_{qj}\Big| \\
&= O(n^{2/3}c_n\log n).
\end{aligned}$$

The above arguments imply that

$$\sqrt{n}H_{1j} = O(n^{1/2}\log^{-1/2} n) \quad \text{for } j = 1,\ldots,p.$$

We complete the proof of (5.1.5).

Step 2.

$$\sqrt{n}H_{2j} = o(n^{1/2}) \text{ for } j = 1,\ldots,p, \quad a.s. \tag{5.1.6}$$

Proof. Observe that

$$\begin{aligned}
\sqrt{n}H_{2j} &= \sum_{i=1}^{n}\Big\{\sum_{k=1}^{n}\widetilde{x}_{kj}\omega_{ni}(T_k)\Big\}\varepsilon_i \\
&= \sum_{i=1}^{n}\Big\{\sum_{k=1}^{n}u_{kj}\omega_{ni}(T_k)\Big\}\varepsilon_i + \sum_{i=1}^{n}\Big\{\sum_{k=1}^{n}h_{nkj}\omega_{ni}(T_k)\Big\}\varepsilon_i \\
&\quad -\sum_{i=1}^{n}\Big[\sum_{k=1}^{n}\Big\{\sum_{q=1}^{n}u_{qj}\omega_{nq}(T_k)\Big\}\omega_{ni}(T_k)\Big]\varepsilon_i.
\end{aligned}$$

We now handle these three terms separately. Applying Lemma A.3 with $a_{ji} = \sum_{k=1}^{n} u_{kj}\omega_{ni}(T_k)$, $r=2$, $V_k = \varepsilon_k$, $1/4 < p_1 < 1/3$ and $p_2 = 1 - p_1$, we obtain that

$$\sum_{i=1}^{n}\Big\{\sum_{k=1}^{n}u_{kj}\omega_{ni}(T_k)\Big\}\varepsilon_i = O(n^{-(2p_1-1)/2}\log n), \quad a.s. \tag{5.1.7}$$

Similarly, by Lemma A.1 and (A.3), we get

$$\begin{aligned}
\Big|\sum_{i=1}^{n}\Big\{\sum_{k=1}^{n}h_{nkj}\omega_{ni}(T_k)\Big\}\varepsilon_i\Big| &\le n\max_{k\le n}\Big|\sum_{i=1}^{n}\omega_{ni}(T_k)\varepsilon_i\Big|\max_{k\le n}|h_{nkj}| \\
&= O(n^{2/3}c_n\log n), \quad a.s. \qquad (5.1.8)
\end{aligned}$$

Analogously, applying (A.3) and Abel's inequality we have

$$\begin{aligned}
\Big|\sum_{i=1}^{n}\Big[\sum_{k=1}^{n}\Big\{\sum_{q=1}^{n}u_{qj}\omega_{nq}(T_k)\Big\}\omega_{ni}(T_k)\Big]\varepsilon_i\Big| &\le n\max_{k\le n}\Big|\sum_{i=1}^{n}\omega_{ni}(T_k)\varepsilon_i\Big|\max_{k\le n}\Big|\sum_{q=1}^{n}u_{qj}\omega_{nq}(T_k)\Big| \\
&= O(n^{1/3}\log^2 n) = o(n^{1/2}), \quad a.s. \qquad (5.1.9)
\end{aligned}$$

A combination of (5.1.7)–(5.1.9) yields (5.1.6).

5.1.3 Appendix

In this subsection we complete the proof of the main result. At first, we state a conclusion, Corollary 5.2.3 of Stout (1974), which will play an elementary role in our proof.

Conclusion S. *Let $V_1, \dots, V_n$ be independent random variables with mean zero. There exists a $\delta_0 > 0$ such that*

$$\max_{1\le i\le n} E|V_i|^{2+\delta_0} < \infty \text{ and } \liminf_{n\to\infty} \frac{1}{n}\sum_{i=1}^n Var(V_i) > 0.$$

Then

$$\limsup_{n\to\infty} \frac{|S_n|}{\sqrt{2s_n^2 \log\log s_n^2}} = 1, \quad a.s.,$$

where $S_n = \sum_{i=1}^n V_i$ and $s_n^2 = \sum_{i=1}^n EV_i^2$.

Recalling that (5.1.3) and (5.1.5) and (5.1.6), in order to complete the proof of (5.1.1), it suffices to prove

$$\limsup_{n\to\infty} \left\{ \frac{|(\Sigma^{-1}\widetilde{\mathbf{X}}\varepsilon)_j|}{2n\log\log n} \right\}^{1/2} = (\sigma^2\sigma^{jj})^{1/2}, \quad a.s. \tag{5.1.10}$$

By a calculation, we get

$$\begin{aligned} \frac{1}{\sqrt{n}}(\Sigma^{-1}\widetilde{\mathbf{X}}\varepsilon)_j &= \sum_{k=1}^p \sigma^{jk}\Big[W_k + \frac{1}{\sqrt{n}}\sum_{q=1}^n \{x_{qk} - h_k(T_q)\}\varepsilon_q\Big] \\ &= \sum_{k=1}^p \sigma^{jk}W_k + \frac{1}{\sqrt{n}}\sum_{q=1}^n \Big(\sum_{k=1}^p \sigma^{jk}u_{qk}\Big)\varepsilon_q, \end{aligned}$$

where

$$W_k = \frac{1}{\sqrt{n}}\sum_{i=1}^n \Big\{h_k(T_i) - \sum_{q=1}^n \omega_{nq}(T_i)x_{qk}\Big\}\varepsilon_i \quad \text{for } 1 \le k \le p.$$

Analogous to the proof of (A.4), by Lemma A.1 we can prove for all $1 \le k \le p$

$$\begin{aligned} |W_k| &\le \frac{1}{\sqrt{n}}\Big|\sum_{i=1}^n \varepsilon_i\Big\{h_k(T_i) - \sum_{q=1}^n \omega_{nq}(T_i)x_{qk}\Big\}\Big| \\ &\quad + \frac{1}{\sqrt{n}}\Big|\sum_{i=1}^n \Big\{\sum_{q=1}^n \omega_{nq}(T_i)u_{qk}\Big\}\Big| \\ &= O(\log n)\cdot o(\log^{-1} n) = o(1) \quad a.s. \end{aligned}$$

using the fact that $\{\varepsilon_i\}$ is independent of (X_i, T_i).

Denote $W_{ij} = \sum_{k=1}^p \sigma^{jk}u_{ik}\varepsilon_i$. Then $EW_{ij} = 0$ for $i = 1, \dots, n$, and

$$E|W_{ij}|^{2+\delta_0} \le C \max_{1\le i\le n} E|\varepsilon_i|^{2+\delta_0} < \infty,$$

and

$$\begin{aligned} \liminf_{n\to\infty} \frac{1}{n}\sum_{i=1}^n EW_{ij}^2 &= \sigma^2(\sigma^{j1}, \dots, \sigma^{jp})\Big(\frac{1}{n}\lim_{n\to\infty}\sum_{i=1}^n u_i u_i^T\Big)(\sigma^{j1}, \dots, \sigma^{jp})^T \\ &= \sigma^2(\sigma^{j1}, \dots, \sigma^{jp})\Sigma(\sigma^{j1}, \dots, \sigma^{jp})^T = \sigma^2\sigma^{jj} > 0. \end{aligned}$$

It follows from Conclusion S that (5.1.10) holds. This completes the proof of (5.1.1).

Next, we prove the second part of Theorem 5.1.1, i.e., (5.1.2). We show that $\sqrt{n}I_i = o(1)$ $(i = 2, 3, 4, 5)$ hold with probability one, and then deal with $\sqrt{n}(I_1 - \sigma^2)$.

It follows from Lemma A.1 and (A.3) that

$$\begin{aligned} |\sqrt{n}I_3| &\leq C\sqrt{n} \max_{1\leq i\leq n}\Big\{\Big|g(T_i) - \sum_{k=1}^{n} \omega_{nk}(T_i)g(T_k)\Big|^2 + \Big|\sum_{k=1}^{n} \omega_{nk}(T_i)\varepsilon_k\Big|^2\Big\} \\ &= o(1), \quad a.s. \end{aligned}$$

It follows from Lemma A.1 and (5.1.5) and (5.1.10) that

$$\sqrt{n}I_2 = o(1), \quad \sqrt{n}I_4 = o(1), \quad a.s.$$

We now consider I_5. We decompose I_5 into three terms and then prove that each term is $o(n^{-1/2})$ with probability one. Precisely,

$$\begin{aligned} I_5 &= \frac{1}{n}\Big\{\sum_{i=1}^{n} \tilde{g}_i\varepsilon_i - \sum_{k=1}^{n} \omega_{ni}(T_k)\varepsilon_k^2 - \sum_{i=1}^{n}\sum_{k\neq i}^{n} \omega_{ni}(T_k)\varepsilon_i\varepsilon_k\Big\} \\ &\stackrel{\text{def}}{=} I_{51} + I_{52} + I_{53}. \end{aligned}$$

From Lemma A.1, we know that

$$\sqrt{n}I_{51} = o(1) \text{ and } \sqrt{n}I_{52} \leq b_n n^{-1/2}\sum_{i=1}^{n} \varepsilon_i^2 = O(\log^{-2} n) = o(1) \quad a.s. \tag{5.1.11}$$

Observe that

$$\sqrt{n}|I_{53}| \leq \frac{1}{\sqrt{n}}\Big|\sum_{i=1}^{n}\sum_{k\neq i} \omega_{nk}(T_i)\varepsilon_i\varepsilon_k - I_n\Big| + I_n \stackrel{\text{def}}{=} \frac{1}{\sqrt{n}}(J_{1n} + I_n),$$

where $I_n = \sum_{i=1}^{n}\sum_{j\neq i}^{n} \omega_{nj}(t_i)(\varepsilon_j' - E\varepsilon_j')(\varepsilon_i' - E\varepsilon_i')$. Similar arguments used in the proof of Lemma A.3 imply that

$$\begin{aligned} J_{1n} &\leq \max_{1\leq j\leq n}\Big|\sum_{i=1}^{n} \omega_{nj}(T_i)\varepsilon_i\Big|\Big\{\sum_{i=1}^{n} |\varepsilon_i''| + E|\varepsilon_i|''\Big\} \\ &\quad + \max_{1\leq j\leq n}\Big|\sum_{i=1}^{n} \omega_{nj}(T_i)(\varepsilon_i' - E\varepsilon_i')\Big|\Big\{\sum_{i=1}^{n} |\varepsilon_i''| + E|\varepsilon_i''|\Big\} \\ &= o(1), \quad a.s. \end{aligned} \tag{5.1.12}$$

It follows from Lemma A.4 and (5.1.12) that

$$\sqrt{n}I_{53} = o(1) \quad a.s. \tag{5.1.13}$$

A combination of (5.1.11)–(5.1.13) leads that $\sqrt{n}I_5 = o(1)$ a.s.

To this end, using Hartman-Winter theorem, we have

$$\limsup_{n\to\infty}\left(\frac{n}{2\log\log n}\right)^{1/2}|I_1-\sigma^2| = (Var\varepsilon_1^2)^{1/2} \quad a.s.$$

This completes the proof of Theorem 5.1.1.

5.2 The Berry-Esseen Bounds

5.2.1 *Introduction and Results*

It is of theoretical and practical interest to characterize the error of approximation in the CLT. This section is concerned with establishing the **Berry-Esseen bounds** for the estimators defined in (1.2.2) and (1.2.4). Our results focus only on the case where ε_i are i.i.d. and (X_i, T_i) are fixed design points. Similar discussions for the random design case can be found in Hong and Cheng (1992b), Gao (1992), Gao, Hong and Liang (1995) and Liang (1994b).

In order to state the main results of this section, we need the following additional assumption.

Assumption 5.2.1

$$\max_{1\le i\le n}\|u_i\| \le B_0 < \infty, \tag{5.2.1}$$

$$\limsup_{n\to\infty} n^{1/2}|v_{jk}-\sigma_{jk}| < \infty \tag{5.2.2}$$

for all $1 \le j,k \le p$, *where* $1/n\sum_{i=1}^n u_iu_i^T = (v_{jk})_{1\le j,k\le p}$ *and* $\Sigma = (\sigma_{jk})_{1\le j,k\le p}$.

Let β_{LSj} and β_j denote the $j-$th components of β_{LS} and β respectively.

Theorem 5.2.1 *Assume that Assumptions 5.2.1 and 1.3.1-1.3.3 hold. Let* $E|\varepsilon_1|^3 < \infty$. *Then for all* $1 \le j \le p$ *and large enough* n

$$\sup_x\left|P\{\sqrt{n}(\beta_{LSj}-\beta_j)/(\sigma^2\sigma^{jj})^{1/2} < x\} - \Phi(x)\right| = O(n^{1/2}c_n^2), \tag{5.2.3}$$

Furthermore, for $c_n = Cn^{-1/2}$ *we have for all* $1 \le j \le p$ *and large enough* n

$$\sup_x\left|P\{\sqrt{n}(\beta_{LSj}-\beta_j)/(\sigma^2\sigma^{jj})^{1/2} < x\} - \Phi(x)\right| = O(n^{-1/2}). \tag{5.2.4}$$

Theorem 5.2.2 *Suppose that Assumptions 5.2.1 and 1.3.1-1.3.3 hold and that* $E\varepsilon_1^6 < \infty$. *Let* $b_n = Cn^{-2/3}$ *in Assumption 1.3.3. Then*

$$\sup_x\left|P\{\sqrt{n}(\widehat{\sigma}_n^2-\sigma^2)/\sqrt{var\varepsilon_1^2} < x\} - \Phi(x)\right| = O(n^{-1/2}). \tag{5.2.5}$$

Remark 5.2.1 *This section mainly establishes the Berry-Esseen bounds for the LS estimator* β_{LS} *and the error estimate* $\widehat{\sigma}_n^2$. *In order to ensure that the two estimators can achieve the optimal Berry-Esseen bound* $n^{-1/2}$, *we assume that the nonparametric weight functions* ω_{ni} *satisfy Assumption 1.3.3 with* $c_n = cn^{-1/2}$. *The restriction of* c_n *is equivalent to selecting* $h_n = cn^{-1/4}$ *in the kernel regression case since* $E\beta_{LS} - \beta = O(h_n^4) + O(h_n^{3/2}n^{-1/2})$ *under Assumptions 1.3.2 and 1.3.3. See also Theorem 2 of Speckman (1988). As mentioned above, the* $n^{-1/5}$ *rate for the bandwidth is required for establishing that the LS estimate* β_{LS} *is* $\sqrt{n}$-*consistent. For this rate of bandwidth, the Berry-Esseen bounds are only of order* $\sqrt{n}h_n^4 = n^{-3/10}$. *In order to establish the optimal Berry-Esseen rate* $n^{-1/2}$, *the faster* $n^{-1/4}$ *rate for the bandwidth is required. This is reasonable.*

5.2.2 Basic Facts

Lemma 5.2.1 *(i) Let* $W_n = U_n + V_n$ *be a sequence of random variables and let* F_n *and* G_n *be the distributions of* W_n *and* U_n, *respectively. Assume that*

$$\|G_n - \Phi\| \le Cn^{-1/2} \quad \text{and} \quad P\{|V_n| \ge Cn^{-1/2}\} \le Cn^{-1/2}.$$

Then $\|F_n - \Phi\| \le Cn^{-1/2}$, *where* $\|F_n - \Phi\| = \sup_x |F_n(x) - \Phi(x)|$.

(ii) Let $W_n = A_nU_n + B_n$, *where* A_n *and* B_n *are real number sequences such that* $|A_n - 1| < Cn^{-1/2}$ *and* $|B_n| < Cn^{-1/2}$. *Assume that* $\|G_n - \Phi\| \le Cn^{-1/2}$. *Then* $\|F_n - \Phi\| \le Cn^{-1/2}$.

Proof. See Lemma 1 of Zhao (1984).

Lemma 5.2.2 *Let* $V_1, \cdots, V_n$ *be i.i.d. random variables. Let* Z_{ni} *be the functions of* V_i *and* Z_{njk} *be the symmetric functions of* V_j *and* V_k. *Assume that* $EZ_{ni} = 0$ *for* $i = 1, \cdots n$, *and* $E(Z_{njk}|V_j) = 0$ *for* $1 \le j \ne k \le n$. *Furthermore* $|C_n| \le Cn^{-3/2}$,

$$D_n^2 = \frac{1}{n}\sum_{i=1}^n EZ_{ni}^2 \le C > 0, \qquad \max_{1\le i\le n} E|Z_{ni}|^3 \le C < \infty,$$

$$E|Z_{njk}|^2 \le d_{nj}^2 + d_{nk}^2, \quad \text{and} \quad \sum_{k=1}^n d_{nk}^2 \le Cn.$$

Putting

$$L_n = \frac{1}{\sqrt{n}D_n}\sum_{i=1}^n Z_{ni} + C_n \sum_{1\le i<k\le n} Z_{njk}.$$

Then

$$\sup_x |P\{L_n \le x\} - \Phi(x)| \le Cn^{-1/2}.$$

Proof. See Theorem 2 of Zhao (1984).

Lemma 5.2.3 *Assume that Assumption 1.3.3 holds. Let $E\varepsilon_1^6 < \infty$. Then*

$$P\Big\{\max_{1\le i\le n}\Big|\sum_{k=1}^n \omega_{nk}(T_i)\varepsilon_k\Big| > C_1 n^{-1/4}\log n\Big\} \le Cn^{-1/2}. \tag{5.2.6}$$

$$P\Big\{\Big|\sum_{i=1}^n\sum_{j\ne i}^n \omega_{nj}(T_i)\varepsilon_j\varepsilon_i\Big| > C_1(n/\log n)^{1/2}\Big\} \le Cn^{-1/2}. \tag{5.2.7}$$

$$P\Big\{\max_{1\le k\le n}\Big|\sum_{i=1}^n\Big(\sum_{j=1}^n \omega_{nk}(T_j)\omega_{ni}(T_j)\Big)\varepsilon_i\Big| > C_1 n^{-1/4}\log n\Big\} \le Cn^{-1/2}. \tag{5.2.8}$$

$$P\Big\{\Big|\sum_{i=1}^n\sum_{j\ne i}^n\Big(\sum_{k=1}^n \omega_{ni}(T_k)\omega_{nj}(T_k)\Big)\varepsilon_j\varepsilon_i\Big| > C_1(n/\log n)^{1/2}\Big\} \le Cn^{-1/2}. \tag{5.2.9}$$

Proof. a) Firstly, let $\varepsilon_j' = \varepsilon_j I_{(|\varepsilon_j|\le n^{1/4})}$ and $\varepsilon_j'' = \varepsilon_j - \varepsilon_j$ for $j = 1, \cdots, n$. By Assumption 1.3.3(ii) and $E\varepsilon_1^6 < \infty$,

$$\begin{aligned} P\Big\{\max_{1\le i\le n}\Big|\sum_{j=1}^n \omega_{nj}(T_i)(\varepsilon_j'' &- E\varepsilon_j'')\Big| > C_1 n^{-1/4}\Big\} \\ &\le P\Big\{\sum_{j=1}^n b_n(|\varepsilon_j''| + E|\varepsilon_j''|) > C_1 n^{-1/4}\Big\} \\ &\le Cb_n n^{1/4}\Big(\sum_{j=1}^n E|\varepsilon_j''|\Big) \le Cb_n E\varepsilon_1^6 \\ &\le Cn^{-1/2}. \end{aligned} \tag{5.2.10}$$

By Bernstein's inequality and Assumption 1.3.3(i)(ii), we have

$$\begin{aligned} P\Big\{\max_{1\le i\le n}\Big|\sum_{j=1}^n \omega_{nj}(T_i)(\varepsilon_j' &- E\varepsilon_j')\Big| > C_1 n^{-1/4}\log n\Big\} \\ &\le \sum_{i=1}^n P\Big\{\Big|\sum_{j=1}^n \omega_{nj}(T_i)(\varepsilon_j' - E\varepsilon_j')\Big| > C_1 n^{-1/4}\log n\Big\} \\ &\le 2\sum_{i=1}^n \exp\Big\{-\frac{C_1 n^{-1/2}\log^2 n}{2\sum_{j=1}^n \omega_{nj}^2(T_i)E\varepsilon_1^2 + 2C_1 b_n \log n}\Big\} \\ &\le 2n\exp\{-C_1^2 C\log n\} \le Cn^{-1/2} \end{aligned} \tag{5.2.11}$$

for large enough C_1. Combining (5.2.10) with (5.2.11), we finish the proof of (5.2.6).

b) Secondly, let

$$I_n = \sum_{i=1}^{n} \sum_{j \neq i}^{n} \omega_{nj}(T_i)(\varepsilon_j' - E\varepsilon_j')(\varepsilon_i' - E\varepsilon_i').$$

Note that

$$\begin{aligned} \Big|\sum_{i=1}^{n} \sum_{j \neq i}^{n} \omega_{ni}(T_i)\varepsilon_i\varepsilon_j - I_n\Big| &\leq \max_{1 \leq i \leq n}\Big|\sum_{j=1}^{n} \omega_{nj}(T_i)\varepsilon_j\Big| \cdot \sum_{j=1}^{n}(|\varepsilon_j''| + E|\varepsilon_j''|) \\ + \max_{1 \leq j \leq n}\Big|\sum_{i=1}^{n} \omega_{nj}(T_i)(\varepsilon_i' &- E\varepsilon_i')\Big| \cdot \sum_{j=1}^{n}(|\varepsilon_j''| + E|\varepsilon_j''|) \\ &\stackrel{\text{def}}{=} J_n. \end{aligned} \quad (5.2.12)$$

By the same reason as (5.2.11),

$$P\Big\{\max_{1 \leq j \leq n}\Big|\sum_{i=1}^{n} \omega_{nj}(T_i)(\varepsilon_i' - E\varepsilon_i')\Big| > C_1 n^{-1/4} \log n\Big\} \leq Cn^{-1/2}. \quad (5.2.13)$$

Next, by $E\varepsilon_1^6 < \infty$ we have

$$\begin{aligned} P\Big\{\sum_{i=1}^{n}(|\varepsilon_i''| + E|\varepsilon_i''|) > C_1 n^{3/4}(\log n)^{-3/2}\Big\} &\leq Cn^{-3/4}(\log n)^{3/2} \sum_{i=1}^{n} E|\varepsilon_i''| \\ &\leq Cn^{1/4}(\log n)^{3/2} E|\varepsilon_1| I_{(|\varepsilon_1| \geq n^{1/4})} \\ &\leq C \cdot n^{-1/2}. \end{aligned} \quad (5.2.14)$$

Thus, by (5.2.6) and (5.2.14) we get

$$P\{J_n \geq C_1(n/\log n)^{1/2}\} \leq Cn^{-1/2}. \quad (5.2.15)$$

On the other hand, by the similar way as for (5.2.11)

$$\max_{1 \leq i \leq n}\Big|\sum_{j=1}^{n} \omega_{nj}(T_i)(\varepsilon_j' - E\varepsilon_j')\Big| = O(n^{-1/4}(\log n)^{-1/2}) \ \ a.s. \quad (5.2.16)$$

In view of (5.2.12)-(5.2.15), in order to prove (5.2.7), it suffices to show that for some suitable C_1

$$P\{|I_n| > C_1(n/\log n)^{1/2}\} < Cn^{-1/2} \quad (5.2.17)$$

whose proof is similar to that of Lemma A.4.

c) The proofs of (5.2.8) and (5.2.9) can be completed by the similar reason as (5.2.6) and (5.2.7).

Lemma 5.2.4 *Assume that Assumptions 1.3.1-1.3.3 and that $E\varepsilon_1^6 < \infty$ hold. Let $(a_1, \cdots, a_n)_j$ denote the j-th element of $(a_1, \cdots, a_n)$. Then for all $1 \le j \le p$*

$$P\{|(\widetilde{\mathbf{X}}^T\varepsilon)_j| > cn^{3/4}(\log n)^{-1/4}\} \le Cn^{-1/2}, \tag{5.2.18}$$

$$P\{|(\widetilde{\mathbf{X}}^T\widehat{\mathbf{G}})_j| > cn^{3/4}(\log n)^{-1/4}\} \le Cn^{-1/2}. \tag{5.2.19}$$

Proof. a) Firstly, observe that

$$(\widetilde{\mathbf{X}}^T\varepsilon)_j = \sum_{i=1}^n u_{ij}\varepsilon_i + \sum_{i=1}^n h_{nij}\varepsilon_i - \sum_{i=1}^n \Big\{\sum_{k=1}^n \omega_{nk}(T_i)u_{kj}\Big\}\varepsilon_i.$$

Using $\max_{1\le i\le n}|u_{ij}| \le C$ and Assumption 1.3.1, and applying **Bernstein's inequality** for all $1 \le j \le p$,

$$\begin{aligned} P\Big\{\Big|\sum_{i=1}^n u_{ij}(\varepsilon_i' \quad &- \quad E\varepsilon_i')\Big| > Cn^{3/4}(\log n)^{-1/4}\Big\} \\ &\le 2\exp\left\{\frac{-Cn^{3/2}(\log n)^{-1/2}}{\sum_{i=1}^n u_{ij}^2 \mathrm{Var}(\varepsilon_i) + n^{1/2}(\log n)^{-1/4}\max_{1\le i\le n}|u_{ij}|}\right\} \\ &\le Cn^{-1/2}. \end{aligned} \tag{5.2.20}$$

Similarly by Assumption 1.3.1,

$$\begin{aligned} P\Big\{\Big|\sum_{i=1}^n u_{ij}(\varepsilon_i'' - E\varepsilon_i'')\Big| &> Cn^{3/4}(\log n)^{-1/4}\Big\} \\ &\le 2n^{-3/2}(\log n)^{1/2}\sum_{i=1}^n u_{ij}^2 E(\varepsilon_i'' - E\varepsilon_i'')^2 \\ &\le Cn^{-2}(\log n)^{1/2}\sum_{i=1}^n u_{ij}^2 E\varepsilon_1^4 \le Cn^{-1/2}. \end{aligned} \tag{5.2.21}$$

Thus, by (5.2.20) and (5.2.21) we get

$$P\Big\{\Big|\sum_{i=1}^n u_{ij}\varepsilon_i\Big| > Cn^{3/4}(\log n)^{-1/4}\Big\} \le Cn^{-1/2}. \tag{5.2.22}$$

Analogous to the proof of (5.2.22), using Assumption 1.3.3(i) and $\max_{1\le i\le n}\|u_i\| \le C$,

$$P\Big\{\Big|\sum_{i=1}^n\sum_{k=1}^n \omega_{nk}(T_i)u_{kj}\varepsilon_i\Big| > Cn^{3/4}(\log n)^{-1/4}\Big\} \le Cn^{-1/2}. \tag{5.2.23}$$

On the other hand, applying Lemma A.3 we have

$$P\Big\{\Big|\sum_{i=1}^n h_{nij}\varepsilon_i\Big| > Cn^{3/4}(\log n)^{-1/4}\Big\} \le Cn^{-1/2}. \tag{5.2.24}$$

Therefore, by (5.2.22)-(5.2.24) we complete the proof of (5.2.18).

b) Secondly, observe that

$$\begin{aligned}
(\widetilde{\mathbf{X}}^T\widehat{\mathbf{G}})_j &= \sum_{i=1}^n\Big\{x_{ij}-\sum_{k=1}^n\omega_{nk}(T_i)x_{kj}\Big\}\{g(T_i)-\widehat{g}_n(T_i)\} \\
&= \sum_{i=1}^n\widetilde{g}_iu_{ij}+\sum_{i=1}^n\widetilde{g}_ih_{nij}-\sum_{k=1}^n\Big\{\sum_{i=1}^n\omega_{nk}(T_i)\widetilde{g}_i\Big\}u_{kj} \\
&\quad -\sum_{k=1}^n\Big\{\sum_{i=1}^n\omega_{nk}(T_i)h_{nij}\Big\}\varepsilon_k-\sum_{k=1}^n\Big\{\sum_{i=1}^n\omega_{nk}(T_i)u_{ij}\Big\}\varepsilon_k \\
&\quad +\sum_{k=1}^n\Big[\sum_{i=1}^n\omega_{nk}(T_i)\Big\{\sum_{q=1}^n\omega_{nq}(T_i)u_{qj}\Big\}\Big]\varepsilon_k \\
&\stackrel{\text{def}}{=} \sum_{k=1}^6 J_{kj}.
\end{aligned} \tag{5.2.25}$$

Applying **Abel's inequality** and using Assumption 1.3.1 and Lemma A.1,

$$J_{kj}=O(a_nc_n) \quad \text{for} \quad k=1,3 \quad \text{and} \quad J_{2j}=O(nc_n^2). \tag{5.2.26}$$

Similar to the proof of (5.2.23), for $k=5,6$

$$P\{|J_{kj}|>Cn^{3/4}(\log n)^{-1/4}\}\le Cn^{-1/2}. \tag{5.2.27}$$

On the other hand, applying Lemma A.3 we have

$$P\{|J_{4j}|>Cn^{3/4}(\log n)^{-1/4}\}\le Cn^{-1/2}. \tag{5.2.28}$$

Combining (5.2.25)-(5.2.28), we complete the proof of (5.2.19).

5.2.3 *Technical Details*

The proof of Theorem 5.2.1

Here we prove only the case of $j=1$, and the others follow by the same reason. Denote $n^{-1}\widetilde{\mathbf{X}}^T\widetilde{\mathbf{X}}=(a_{jk})_{1\le j,k\le p}$, $n(\widetilde{\mathbf{X}}^T\widetilde{\mathbf{X}})^{-1}=(a^{jk})_{1\le j,k\le p}$,

$$\begin{aligned}
p_{ni} &= \sum_{j=1}^p a^{1j}\Big\{x_{ij}-\sum_{k=1}^n\omega_{nk}(T_i)x_{kj}\Big\}=\sum_{j=1}^p a^{1j}\widetilde{x}_{ij}, \\
q_{ni} &= p_{ni}-\sum_{k=1}^n\omega_{nk}(T_i)p_{nk}.
\end{aligned} \tag{5.2.29}$$

Then, by Assumption 1.3.1 we have

$$n^{1/2}(\widehat{\beta}_{n1}-\beta_1)=n^{-1/2}\sum_{i=1}^n q_{ni}\varepsilon_i+n^{1/2}\sum_{i=1}^n p_{ni}\widetilde{g}_i. \tag{5.2.30}$$

Noting that Assumptions 1.3.1, 1.3.2 and 1.3.3(i) and

$$\max_{1\le i\le n}|x_{ij}| \le \max_{1\le i\le n}|h_j(T_i)| + \max_{1\le i\le n}|u_{ij}| \le C < \infty,$$

we obtain for large enough n

$$\max_{1\le i\le n}|p_{ni}| \le C \quad \text{and} \quad \max_{1\le i\le n}|q_{ni}| \le C.$$

Let $Z_{ni} = q_{ni}\varepsilon_i$, then $EZ_{ni} = 0$, and

$$D_n^2 = n^{-1}\sum_{i=1}^n EZ_{ni}^2 = n^{-1}\sigma^2\sum_{i=1}^n q_{ni}^2 \ge C > 0,$$
$$\max_{1\le i\le n} E|Z_{ni}|^3 = \max_{1\le i\le n}|q_{ni}|^3 E|\varepsilon_1|^3 \le C < \infty.$$

It follows from (5.2.39) that

$$D_n^2 = n^{-1}\sigma^2\sum_{i=1}^n q_{ni}^2 \ge C > 0 \quad \text{for large enough } n$$

Therefore, by Lemma 5.2.2 we obtain

$$\sup_x\Big|P\Big\{n^{-1/2}D_n^{-1}\sum_{i=1}^n Z_{ni} < x\Big\} - \Phi(x)\Big| \le Cn^{-1/2}. \tag{5.2.31}$$

In the following, we need to determine the order of D_n^2.

Let $a^j = (a^{j1},\cdots,a^{jp})^T$ and $\sigma^j = (\sigma^{j1},\cdots,\sigma^{jp})^T$ for $1 \le j \le p$. By Lemma A.2(i) we get

$$\begin{aligned}\lim_{n\to\infty} n^{-1}\sum_{i=1}^n p_{ni}^2 &= \lim_{n\to\infty} n^{-1}(a^1)^T\sum_{i=1}^n \widetilde{x}_i\widetilde{x}_i^T\cdot(a^1)\\ &= (\sigma^1)^T\cdot\Sigma\cdot(\sigma^1) = \sigma^{11}.\end{aligned} \tag{5.2.32}$$

We now prove

$$\sum_{i=1}^n\Big\{\sum_{j=1}^n \omega_{ni}(T_j)p_{nj}\Big\}^2 = O(n^{1/2}). \tag{5.2.33}$$

Denote

$$m_{ks} = \sum_{i=1}^n \omega_{ni}(T_k)\omega_{ni}(T_s), \quad 1 \le k \ne s \le n, \quad m_s = \sum_{k=1}^n m_{ks}p_{nk}.$$

By `Abel's inequality`, we obtain

$$\Big|\sum_{i=1}^n p_{ni}m_i\Big| \le 6\max_{1\le k\le n}\Big|\sum_{i=1}^k p_{ni}\Big|\cdot\max_{1\le i\le n}|m_i|, \tag{5.2.34}$$

$$|m_s| = \Big|\sum_{i=1}^n p_{ni}m_{is}\Big| \le 6\max_{1\le k\le n}\Big|\sum_{i=1}^k p_{ni}\Big|\cdot\max_{1\le k,s\le n}\omega_{nk}(T_s) \tag{5.2.35}$$

and

$$\max_{1\le k\le n}\Big|\sum_{s=1}^{n}\Big\{\sum_{i=1}^{k}\omega_{ns}(T_i)\Big\}u_{sj}\Big| = O(a_n). \tag{5.2.36}$$

Also, by Assumption 1.3.1 and Lemma A.1, we have

$$\begin{aligned}
\max_{1\le k\le n}\Big|\sum_{i=1}^{k}p_{ni}\Big| &\le \max_{1\le k\le n}\Big|\sum_{j=1}^{p}\sum_{i=1}^{k}\widetilde{x}_{ij}a^{1j}\Big| \le \max_{1\le k\le n}\Big|\sum_{j=1}^{p}\sum_{i=1}^{k}u_{ij}a^{1j}\Big| \\
&\quad + \max_{1\le k\le n}\Big|\sum_{j=1}^{p}\Big[\sum_{i=1}^{k}\sum_{s=1}^{n}\omega_{ns}(T_i)\{h_j(T_s)-h_j(T_i)\}\Big]a^{1j}\Big| \\
&\quad + \max_{1\le k\le n}\Big|\sum_{j=1}^{p}\Big\{\sum_{i=1}^{k}\sum_{s=1}^{n}\omega_{ns}(T_i)u_{sj}\Big\}a^{1j}\Big| \\
&= O(a_n)+O(nc_n).
\end{aligned} \tag{5.2.37}$$

Thus, by Assumption 1.3.3, (5.2.32) and (5.2.34)-(5.2.37), we obtain

$$\sum_{s=1}^{n}\Big[\sum_{k\ne s}^{n}\Big\{\sum_{i=1}^{n}\omega_{ni}(T_k)\omega_{ni}(T_s)\Big\}p_{nk}\Big]p_{ns} = O(a_n^2b_n)+O(n^2b_nc_n^2)$$

and

$$\begin{aligned}
\sum_{i=1}^{n}\Big\{\sum_{k=1}^{n}\omega_{ni}(T_k)p_{nk}\Big\}^2 &= \sum_{s=1}^{n}\sum_{k=1}^{n}\omega_{ni}^2(T_k)p_{nk}^2+\sum_{i=1}^{n}\sum_{k\ne s}^{n}\omega_{ni}(T_k)\omega_{ni}(T_s)p_{nk}p_{ns} \\
&= O(nb_n)+O(a_n^2b_n)+O(n^2b_nc_n^2) \\
&= O(n^{3/2}c_n).
\end{aligned} \tag{5.2.38}$$

On the other hand, by (5.2.32), (5.2.38) and the definition of q_{ni}, we obtain

$$\lim_{n\to\infty}n^{-1}\sum_{i=1}^{n}q_{ni}^2 = \lim_{n\to\infty}n^{-1}\sum_{i=1}^{n}p_{ni}^2 = \sigma^{11}>0. \tag{5.2.39}$$

Furthermore, by (5.2.29) and Lemma A.2(ii) we have for all (j,k)

$$|a_{jk}-\sigma_{jk}|\le Cn^{-1/2} \quad \text{and} \quad \Big|n^{-1}\sum_{i=1}p_{ni}^2-\sigma^{11}\Big|\le Cn^{-1/2}.$$

Also, by (5.2.35), (5.2.38), and Assumption 1.3.3(ii) and using the similar reason as (5.2.34),

$$\begin{aligned}
\Big|n^{-1}\sum_{i=1}^{n}q_{ni}^2-\sigma^{11}\Big| &\le \Big|n^{-1}\sum_{i=1}^{n}p_{ni}^2-\sigma^{11}\Big|+n^{-1}\sum_{i=1}^{n}\Big\{\sum_{k=1}^{n}\omega_{ni}(T_k)p_{nk}\Big\}^2 \\
&\quad +12n^{-1}\max_{1\le k\le n}\Big|\sum_{i=1}^{k}p_{ni}\Big|\cdot\max_{1\le i\le n}\Big|\sum_{k=1}^{n}\omega_{ni}(T_k)p_{nk}\Big| \\
&\le Cn^{1/2}c_n^2.
\end{aligned} \tag{5.2.40}$$

By the similar reason as in the proof of (5.2.34), using (5.2.37), Lemma A.1, and Assumption 1.3.3, we obtain

$$\Big|\sum_{i=1}^{n} p_{ni}\widetilde{g}_i\Big| \leq 6 \max_{1\leq k\leq n}\Big|\sum_{i=1}^{k} p_{ni}\Big| \max_{1\leq i\leq n} |\widetilde{g}_i| \leq Cnc_n^2. \tag{5.2.41}$$

Therefore, by (5.2.30), (5.2.31), (5.2.40) and (5.2.41), and using the conditions of Theorem 5.2.1, we have

$$\sup_x \Big|P\{\sqrt{n}(\widehat{\beta}_{LS1} - \beta_1)/(\sigma^2\sigma^{11})^{1/2} < x\} - \Phi(x)\Big| = O(n^{1/2}c_n^2).$$

Thus, we complete the proof of (5.2.3). The proof of (5.2.4) follows similarly.

The proof of Theorem 5.2.2

In view of Lemmas 5.2.1 and 5.2.2 and the expression given in (5.1.4), in order to prove Theorem 5.2.2, we only need to prove $P\{\sqrt{n}|I_k| > Cn^{-1/2}\} < Cn^{-1/2}$ for $k = 2, 3, 4, 5$ and large enough $C > 0$, which are arranged in the following lemmas.

Lemma 5.2.5 *Assume that Assumptions 1.3.1-1.3.3 hold. Let $E\varepsilon_1^6 < \infty$. Then*

$$P\{|\sqrt{n}I_2| > Cn^{-1/2}\} \leq Cn^{-1/2}.$$

Proof. According to the definition of I_2, it suffices to show that

$$P\Big\{\frac{1}{\sqrt{n}}\varepsilon^T\mathbf{U}(\mathbf{U}^T\mathbf{U})^{-1}\mathbf{U}^T\varepsilon| > C_1 n^{-1/2}\Big\} \leq \frac{C_2}{\sqrt{n}}$$

for some constants $C_1 > pE\varepsilon_1^2$ and $C_2 > 0$. Let $\{p_{ij}\}$ denote the (i,j) element of the matrix $\mathbf{U}(\mathbf{U}^T\mathbf{U})^{-1}\mathbf{U}^T$. We now have

$$\begin{aligned}
P\Big\{\varepsilon^T\mathbf{U}(\mathbf{U}^T\mathbf{U})^{-1}\mathbf{U}^T\varepsilon > C_1\Big\} &= P\Big\{\sum_{i=1}^{n} p_{ii}\varepsilon_i^2 + \sum_{i=1}^{n}\sum_{j\neq i} p_{ij}\varepsilon_i\varepsilon_j > C_1\Big\} \\
&= P\Big\{\sum_{i=1}^{n} p_{ii}(\varepsilon_i^2 - E\varepsilon_i^2) \\
&\qquad + \sum_{i=1}^{n}\sum_{j\neq i} p_{ij}\varepsilon_i\varepsilon_j > C_1 - pE\varepsilon_1^2\Big\} \\
&\leq P\Big\{\Big|\sum_{i=1}^{n} p_{ii}(\varepsilon_i^2 - E\varepsilon_i^2)\Big| > \frac{1}{2}(C_1 - pE\varepsilon_1^2)\Big\} \\
&\quad + P\Big\{\Big|\sum_{i=1}^{n}\sum_{j\neq i} p_{ij}\varepsilon_i\varepsilon_j\Big| > \frac{1}{2}(C_1 - pE\varepsilon_1^2)\Big\} \\
&\stackrel{\text{def}}{=} I_{21} + I_{22}.
\end{aligned}$$

Using (5.2.1) we have

$$I_{21} \leq \frac{1}{C_2^2} E\left\{\sum_{i=1}^n p_{ii}(\varepsilon_i^2 - E\varepsilon_i^2)\right\}^2 \leq \max_{1\leq i\leq n} p_{ii}\frac{p}{C_2^2} E(\varepsilon_i^2 - E\varepsilon_i^2)^2 \leq \frac{C_3}{n},$$

where $C_2 = \frac{1}{2}(C_1 - pE\varepsilon_1^2)$ and $C_3 > 0$ is a constant.

Similarly, we can show that $I_{22} \leq C/\sqrt{n}$.

Lemma 5.2.6 *Assume that Assumptions 1.3.1 and 1.3.3 hold. Let $E\varepsilon_1^6 < \infty$. Then $P\{|\sqrt{n}I_3| > Cn^{-1/2}\} \leq Cn^{-1/2}$.*

Proof. In view of Lemma A.1 and the definition of $\widehat{\mathbf{G}}$, in order to prove Lemma 5.2.6, it suffices to show that

$$P\left\{\sum_{i=1}^n \left(\sum_{k=1}^n W_{nk}(T_i)\varepsilon_k\right)^2 > C_1\right\} \leq \frac{C_2}{\sqrt{n}}$$

for some positive constant C_1 and C_2. Observe that

$$\begin{aligned}\sum_{i=1}^n \left(\sum_{k=1}^n W_{nk}(T_i)\varepsilon_k\right)^2 &= \sum_{k=1}^n \left(\sum_{i=1}^n W_{nk}^2(T_i)\right)(\varepsilon_k - E\varepsilon_k^2) \\ &+ \sum_{i=1}^n\sum_{k=1}^n\sum_{l\neq k}^n W_{nk}(T_i)W_{nl}(T_i)\varepsilon_k\varepsilon_l + E\varepsilon_1^2\sum_{k=1}^n\sum_{i=1}^n W_{nk}^2(T_i).\end{aligned}$$

Thus we can finish the proof of Lemma 5.2.6 by using the similar reason as the proofs of Lemmas 5.2.3 and 5.2.5.

Lemma 5.2.7 *Assume that Assumptions 1.3.1-1.3.3 hold. Let $E\varepsilon_1^6 < \infty$. Then*

$$P\{|\sqrt{n}I_4| > Cn^{-1/2}\} \leq Cn^{-1/2}. \tag{5.2.42}$$

Proof. The proof is similar to that of Lemma 5.2.5. We omit the details.

We now handle I_5. Observe the following decomposition

$$\frac{1}{n}\Big\{\sum_{i=1}^n \widetilde{g}_i\varepsilon_i - \sum_{k=1}^n \omega_{nk}(T_k)\varepsilon_k^2 - \sum_{i=1}^n\sum_{k\neq i}^n \omega_{ni}(T_k)\varepsilon_i\varepsilon_k\Big\} \stackrel{\text{def}}{=} I_{51} + I_{52} + I_{53}. \tag{5.2.43}$$

For $q = 1/4$, we have

$$\begin{aligned}P\{\sqrt{n}|I_{51}| > Cn^{-1/2}\} &\leq P\Big\{\Big|\sum_{i=1}^n \widetilde{g}_i\varepsilon_i I_{(|\varepsilon_i|\leq n^q)}\Big| > C\Big\} + P\Big\{\Big|\sum_{i=1}^n \widetilde{g}_i\varepsilon_i I_{(|\varepsilon_i|> n^q)}\Big| > C\Big\} \\ &\stackrel{\text{def}}{=} R_{1n} + R_{2n}.\end{aligned}$$

It follows from `Bernstein inequality` and Lemma A.1 that

$$R_{1n} \leq 2\exp\{-Cn^{1/2-q}\} < Cn^{-1/2}$$

and

$$\begin{aligned} R_{2n} &\leq C\sum_{i=1}^{n}|\tilde{g}_i|\cdot E|\varepsilon_i I_{(|\varepsilon_i|>n^q)}| \\ &\leq C\max_{1\leq i\leq n}|\tilde{g}_i|\sum_{i=1}^{n}E\varepsilon_i^6 n^{-6q} < Cn^{-1/2}. \end{aligned}$$

Combining the conclusions for R_{1n} and R_{2n}, we obtain

$$P\{\sqrt{n}|I_{51}| > Cn^{-1/2}\} < Cn^{-1/2}. \tag{5.2.44}$$

By Assumptions 1.3.3 (ii) (iii), we have for any constant $C > C_0E\varepsilon_1^2$

$$\begin{aligned} P\Big\{\sqrt{n}|I_{52}| > Cn^{-1/2}\Big\} &= P\Big\{\Big|\sum_{i=1}^{n}\omega_{ni}(T_i)(\varepsilon_i^2-E\varepsilon_i^2)+\sum_{i=1}^{n}\omega_{ni}(T_i)E\varepsilon_i^2\Big| > C\Big\} \\ &\leq P\Big\{\Big|\sum_{i=1}^{n}\omega_{ni}(T_i)(\varepsilon_i^2-E\varepsilon_i^2)\Big| > C-C_0E\varepsilon_1^2\Big\} \\ &\leq C_1E\Big\{\sum_{i=1}^{n}\omega_{ni}(T_i)(\varepsilon_i^2-E\varepsilon_i^2)\Big\}^2 \\ &\leq C_1\max_{1\leq i\leq n}\omega_{ni}(T_i)\sum_{i=1}^{n}\omega_{ni}(T_i)E(\varepsilon_i^2-E\varepsilon_i^2)^2 \\ &< C_1n^{-1/2}, \end{aligned} \tag{5.2.45}$$

where C_0 satisfies $\max_{n\geq 1}\sum_{i=1}^{n}\omega_{ni}(T_i)\leq C_0$, $C_1=(C-C_0E\varepsilon_1^2)^{-2}$ and C_2 is a positive constant.

Calculate the `tail probability` of I_{53}.

$$\begin{aligned} P\left\{\sqrt{n}|I_{53}| > Cn^{-1/2}\right\} &\leq P\Big\{\Big|\sum_{i=1}^{n}\sum_{k\neq i}\omega_{nk}(T_i)\varepsilon_i\varepsilon_k - I_n\Big| > C\Big\} + P\{|I_n| > C\} \\ &\stackrel{\text{def}}{=} J_{1n}+J_{2n}, \end{aligned}$$

where $I_n=\sum_{i=1}^{n}\sum_{j\neq i}^{n}\omega_{nj}(T_i)(\varepsilon_j'-E\varepsilon_j')(\varepsilon_i'-E\varepsilon_i')$ and $\varepsilon_i'=\varepsilon_iI_{(|\varepsilon_i|\leq n^q)}$.

By a direct calculation, we have

$$\begin{aligned} J_{1n} &\leq P\left\{\max_{1\leq j\leq n}\Big|\sum_{i=1}^{n}\omega_{nj}(T_i)\varepsilon_i\Big|\Big(\sum_{i=1}^{n}|\varepsilon_i''|+E|\varepsilon_i|''\Big) > C\right\} \\ &\quad +P\left\{\max_{1\leq j\leq n}\Big|\sum_{i=1}^{n}\omega_{nj}(T_i)(\varepsilon_i'-E\varepsilon_i')\Big|\Big(\sum_{i=1}^{n}|\varepsilon_i''|+E|\varepsilon_i''|\Big) > C\right\} \\ &\stackrel{\text{def}}{=} J_{1n1}+J_{1n2}. \end{aligned}$$

Similar to the proof of (5.2.11), we have

$$P\Big\{\max_{1\leq j\leq n}\Big|\sum_{i=1}^{n}\omega_{nj}(T_i)(\varepsilon_i' \quad - \quad E\varepsilon_i')\Big| > Cn^{-1/4}\Big\}$$

$$\begin{aligned} &\leq 2\sum_{j=1}^{n} P\Big\{\Big|\sum_{i=1}^{n}\omega_{nj}(T_i)(\varepsilon_i' - E\varepsilon_i')\Big| > Cn^{-1/4}\Big\} \\ &\leq 2\sum_{j=1}^{n}\exp\Big(-\frac{cn^{-1/2}}{2\sum_{i=1}^{n}\omega_{nj}^2(T_i)E\varepsilon_1^2 + 2Cb_n}\Big) \\ &\leq 2n\exp(-Cn^{-1/2}b_n^{-1}) \leq Cn^{-1/2}. \end{aligned} \tag{5.2.46}$$

On the other hand, applying **Chebyshev's inequality**,

$$\begin{aligned} P\Big\{\sum_{i=1}^{n}|\varepsilon_i''| + E|\varepsilon_i''| > Cn^{1/4}\Big\} &\leq Cn^{-1/4}E\sum_{i=1}^{n}|\varepsilon_i''| \\ &\leq Cn^{-1/4}\sum_{i=1}^{n}E\varepsilon_i^6 n^{-5/4} < Cn^{-1/2}. \end{aligned} \tag{5.2.47}$$

Hence, (5.2.46) and (5.2.47) imply

$$J_{1n2} < Cn^{-1/2}. \tag{5.2.48}$$

Similar to (5.2.46) and (5.2.47) we have

$$J_{1n1} < Cn^{-1/2}. \tag{5.2.49}$$

Combining (5.2.48) with (5.2.49), we obtain

$$J_{1n} < Cn^{-1/2}. \tag{5.2.50}$$

Thus, from (5.2.50) and Lemma A.4, we get

$$P\{\sqrt{n}|I_{53}| > Cn^{-1/2}\} < Cn^{-1/2}. \tag{5.2.51}$$

It follows from (5.2.43), (5.2.44), (5.2.45) and (5.2.51) that

$$P\{\sqrt{n}|I_5| > Cn^{-1/2}\} < Cn^{-1/2}. \tag{5.2.52}$$

The proof of Theorem 5.2.2. From Lemmas 5.2.5, 5.2.6, 5.2.7 and (5.2.52), we have

$$P\left\{\sqrt{\frac{n}{\mathrm{Var}(\varepsilon_1^2)}}\Big|-I_2 + I_3 - 2I_4 + 2I_5\Big| > Cn^{-1/2}\right\} < Cn^{-1/2}. \tag{5.2.53}$$

Let $Z_{ni} = \varepsilon_i^2 - \sigma^2$ and $Z_{nij} = 0$. Then $D_n^2 = 1/n\sum_{i=1}^{n}EZ_{ni}^2 = Var(\varepsilon_1^2)$ and $E|Z_{ni}|^3 \leq E|\varepsilon_1^2 - \sigma^2|^3 < \infty$, which and Lemma 5.2.2 imply that

$$\sup_x |P\{\sqrt{n}D_n^{-1}(I_1 - \sigma^2) < x\} - \Phi(x)| < Cn^{-1/2}. \tag{5.2.54}$$

Applying Lemma 5.2.1, (5.2.53) and (5.2.54), we complete the proof of Theorem 5.2.2.

5.3 Asymptotically Efficient Estimation

5.3.1 Motivation

Recently, Cuzick (1992a) constructed **asymptotically efficient** estimators for β for the case where the error density is known. The problem was extended later to the case of unknown error distribution by Cuzick (1992b) and Schick (1993). Golubev and Härdle (1997), under the assumption that X and T are mutually independent, investigated the **second order risk**, and calculated it exactly up to constant. Their further study shows that the spline estimator (Heckman (1986) and Rice (1986)) is not a **second order minimax estimator**.

In this section, we shall consider the case where ε_i are i.i.d. and (X_i, T_i) are random design points, and construct asymptotically efficient estimators in the sense of normality, i.e., their asymptotically variance reaches an optimum value.

Our construction approach is essentially based on the following fact (Cheng, Chen, Chen and Wu 1985, p256): *Let $W_1, \ldots, W_n$ be i.i.d. random variables drawn from the density function $s(w, \theta)$. Set*

$$Z(w,\theta) = \frac{\partial}{\partial\theta}\log s(w,\theta) \text{ and } B_n(\theta) = \frac{1}{n}\sum_{i=1}^{n}\frac{\partial Z(W_i,\theta)}{\partial\theta}$$

If θ_n is a $n^{\alpha}(\alpha > 1/4)$-order consistent estimator of θ, then under appropriate regularity conditions,

$$\theta_{Mn} = \theta_n - \frac{1}{n}B_n^{-1}(\theta_n)\sum_{i=1}^{n} Z(W_i,\theta_n)$$

is an **asymptotically efficient** *estimator of θ.* See also Stone (1975) and Schick (1986).

5.3.2 Construction of Asymptotically Efficient Estimators

The efficiency criterion we shall be using in this section is a **least dispersed regular** estimator as elaborated in Begun, Hall, Huang and Wellner (1983). See Bickel, Klaasen, Ritov and Wellner (1993).

In order to derive **asymptotically efficient** estimators of β, we assume that there exist a root-n rate estimator $\widehat{\beta}$ of β and a $n^{-1/3}\log n$-nonparametric estimator $\widehat{g}(t)$ of $g(t)$. See, for example, the estimators given in (1.2.2) and (1.2.3). We assume that the random error ε has a density function $\varphi(\cdot)$ which has finite

Fisher information

$$I = \int \frac{\varphi'^2}{\varphi}(y)\,dy < \infty.$$

The covariance of the **asymptotically efficient** estimator of β is $\Sigma^{-1}I^{-1}$. See Cuzick (1992b) and Golubev and Härdle (1997). The common root-n rate estimator is asymptotically normal with covariance matrix $\Sigma^{-1}\sigma^2$. Liang (1994a) showed that a root-n asymptotically normal estimator of β is **asymptotically efficient** if and only if ε is Gaussian.

In constructing our estimation procedure, we smooth the density function φ to ensure that it is continuously differentiable, and truncate the integration and use the **split-sample** technique.

Let $L(y)$ be the likelihood function of $\varphi(y)$, i.e., $L(y) = \varphi'/\varphi(y)$. Let $\psi_r(z, r)$ be the density function of the normal distribution $N(0, r^{-2})$. Take $r_n = \log n$ and $f(z, r) = \int \psi_r(z - y, r)\varphi(y)\,dy$. Obviously $f(z, r)$ is continuously differentiable of all orders and converges to $\varphi(z)$ as r tends to infinity. Let $f'(z, r)$ denote the first derivative of $f(z, r)$ on z. $L(z, r) = [f'(z, r)/f(z, r)]_r$, where $[a]_r = a$ if $|a| \le r$, $-r$ if $a \le -r$ and r if $a > r$. $L(z, r)$ is also smoothed as

$$L_n(y, r_n) = \int_{-r_n}^{r_n} L(z, r_n)\psi_{r_n}(y - z, r_n)\,dz.$$

Assume that $\widehat{\beta}_1$, $\widehat{g}_1(t)$ are the estimators of β, $g(t)$ based on the first-half samples $(Y_1, X_1, T_1), \cdots, (Y_{k_n}, X_{k_n}, T_{k_n})$, respectively, while $\widehat{\beta}_2$, $\widehat{g}_2(t)$ are the ones based on the second-half samples $(Y_{k_n+1}, X_{k_n+1}, T_{k_n+1}), \cdots, (Y_n, X_n, T_n)$, respectively, where $k_n (\le n/2)$ is the largest integer part of $n/2$. For the sake of simplicity in notation, we denote $\Lambda_j = Y_j - X_j^T\beta - g(T_j)$, $\Lambda_{n1j} = Y_j - X_j^T\widehat{\beta}_1 - \widehat{g}_1(T_j)$, and $\Lambda_{n2j} = Y_j - X_j^T\widehat{\beta}_2 - \widehat{g}_2(T_j)$.

5.3.2.1 <u>When the Error Density Function φ Known</u>

We construct an estimator of β as follows:

$$\beta_n^* = \bar{\widehat{\beta}} - \frac{1}{n}\Sigma^{-1}I^{-1}\Big\{\sum_{j=1}^{k_n} X_j L_n(\Lambda_{n2j}, r_n) + \sum_{j=k_n+1}^{n} X_j L_n(\Lambda_{n1j}, r_n)\Big\},$$

where $\bar{\widehat{\beta}}$, $\bar{\widehat{\beta}}_1$, $\bar{\widehat{\beta}}_2$ are discretized forms of $\widehat{\beta}$, $\widehat{\beta}_1$, $\widehat{\beta}_2$ respectively.

Theorem 5.3.1 *Suppose that Σ is a positive definite matrix and that $\widehat{\beta}$, $\widehat{\beta}_1$, $\widehat{\beta}_2$; $\widehat{g}(t)$, $\widehat{g}_1(t)$ and $\widehat{g}_2(t)$ are root-n rate and $n^{-1/3}\log n$ estimators of β and $g(t)$, respectively. Then as $n \to \infty$*

$$\sqrt{n}(\beta_n^* - \beta) \longrightarrow^{\mathcal{L}} N(0, I^{-1}\Sigma^{-1}).$$

5.3.2.2 When the Error Density Function φ Unknown

In this case, β_n^* is not a statistic any more since $L(z, r_n)$ contains the unknown function φ and I is also unknown. We estimate them as follows. Set

$$\begin{aligned}
\widetilde{f}_n(z,r) &= \frac{1}{n}\Big\{\sum_{j=1}^{k_n} \psi_{r_n}(z - \Lambda_{n2j}, r) + \sum_{j=k_n+1}^{n} \psi_{r_n}(z - \Lambda_{n1j}, r)\Big\}, \\
\widetilde{f}_n'(z,r) &= \frac{\partial \widetilde{f}_n(z,r)}{\partial z}, \\
\widetilde{L}_n(z,r_n) &= \Big[\frac{\widetilde{f}_n'(z,r_n)}{\widetilde{f}_n(z,r_n)}\Big]_{r_n}, \quad \widetilde{A}_n(r_n) = \int_{-r_n}^{r_n} \widetilde{L}_n^2(z,r_n)\widetilde{f}_n(z,r_n)\,dz.
\end{aligned}$$

Define

$$\begin{aligned}
\widetilde{\beta}_n = \widetilde{\widetilde{\beta}} \;-\; & \frac{\Sigma^{-1}}{n\widetilde{A}_n(r_n)} \int_{-r_n}^{r_n} \widetilde{L}_n(z,r_n)\Big\{\sum_{j=1}^{k_n} X_j \psi_{r_n}(z - \Lambda_{n2j}, r_n) \\
& + \sum_{j=k_n+1}^{n} X_j \psi_{r_n}(z - \Lambda_{n1j}, r_n)\Big\} dz
\end{aligned}$$

as the estimator of β.

Theorem 5.3.2 *Under the condition of Theorem 5.3.1, we have as $n \to \infty$*

$$\sqrt{n}(\widetilde{\beta}_n - \beta) \longrightarrow^{\mathcal{L}} N(0, I^{-1}\Sigma^{-1}).$$

Remark 5.3.1 *Theorems 5.3.1 and 5.3.2 show us that the estimators β_n^* and $\widetilde{\beta}_n$ are both asymptotically efficient.*

Remark 5.3.2 *Generally Σ is unknown and must be estimated. To do so, let $\widehat{g}_{x,h}(\cdot)$ be the kernel regression estimator of $E(X|T)$. Then*

$$\frac{1}{n}\sum_{i=1}^{n}\{X_i - \widehat{g}_{x,h}(T_i)\}\{X_i - \widehat{g}_{x,h}(T_i)\}^T$$

is a consistent estimator of Σ.

5.3.3 Four Lemmas

In the following Lemmas 5.3.1 and 5.3.2, a_1 and ν are constants satisfying $0 < a_1 < 1/12$ and $\nu = 0, 1, 2$.

Lemma 5.3.1 *As n sufficiently large,*

$$|\psi_{r_n}^{(\nu)}(x, r_n)| < C_\nu r_n^{2+2\nu}\{\psi_{r_n}(x, r_n) + \psi_{r_n}^{1-a_1}(x, r_n)\}$$

uniformly on x, where C_ν is a positive constant depending only on a_1 and ν.

Proof. Let n be large enough such that $a_1 r_n^2 > 1$. Then $\exp(x^2 a_1 r_n^2/2) > |x|^\nu$ if $|x|^\nu > r_n^{a_1}$, and $|x|^\nu \exp(-x^2 r_n^2/2) \le \exp\{-(1-a_1)x^2 r_n^2/2\}$ holds. If $|x|^\nu \le r_n^{a_1}$, then $|x|^\nu \exp(-x^2 r_n^2/2) \le r_n^{a_1} \exp(-x^2 r_n^2/2)$. These arguments deduce that

$$|x|^\nu \exp(-x^2 r_n^2/2) \le r_n^{a_1}\left[\exp\{-(1-a_1)x^2 r_n^2/2\} + \exp(-x^2 r_n^2/2)\right]$$

holds uniformly on x. The proof of this lemma immediately follows.

Lemma 5.3.2 *Let M be a given positive constant. Then*

$$\sup_{|t| \le M n^{-1/4}} \psi_{r_n}(x+t, r_n) \le 3\{\psi_{r_n}(x, r_n) + \psi_{r_n}^{1-a_1}(x, r_n)\}$$

holds uniformly on x as n sufficiently large.

Proof. Let n be large enough such that

$$n^{1/4} r_n^4 > 2Ma_1 \log r_n. \tag{5.3.1}$$

Denote $a_n = r_n^{-2} n^{1/4} M^{-1}$. $|x| < a_n$ implies $|x r_n^2 t| < 1$, and so

$$\left|\frac{\psi_{r_n}(x+t, r_n)}{\psi_{r_n}(x, r_n)}\right| = \exp\{-(2xt + t^2) r_n/2\} = \exp(-xtr_n^2) \le e.$$

When $|x| \ge a_n$, it follows from (5.3.1) that $|x| > 2r_n^2 a_1 \log r_n$. A direct calculation implies

$$\frac{\psi_{r_n}(x+t, r_n)}{\psi_{r_n}^{1-a_1}(x, r_n)} < 1 < e.$$

The proof of the lemma is immediately derived.

Lemma 5.3.3

$$(i) \quad \sup_y |L_n^{(\nu)}(y, r_n)| = O(r_n^{\nu+1});$$

$$(ii) \quad \lim_{n\to\infty} \int_0^1 \int \{L_n(y - v_n(t), r_n) - L_n(y, r_n)\}^2 \varphi(y) h(t)\, dy\, dt = 0 \quad (5.3.2)$$

$$\textit{if} \int_0^1 v_n^2(t) h(t)\, dt = O(n^{-2/3} \log^{2/3} n).$$

Proof. By the definition of $L_n^{(\nu)}(y, r_n)$, we obtain

$$r_n \int_{-r_n}^{r_n} |\psi_{r_n}^{(\nu)}(y - z, r_n)|\, dz \leq r_n \int_{-\infty}^{\infty} |\psi_{r_n}^{(\nu)}(y - z, r_n)|\, dz = O(r_n^{\nu+1}).$$

We complete the proof of (i).

Applying the Taylor expansion, $L_n(y - v_n(t), r_n) - L_n(y, r_n) = L_n'(\tilde{y}_n, r_n) v_n(t)$, where $\tilde{y}_n$ is between y and $y - v_n(t)$. The left-hand side of (5.3.2) equals

$$\lim_{n\to\infty} \int_0^1 \int \{L_n'(\tilde{y}_n, r_n) v_n(t)\}^2 \varphi(y) h(t)\, dy\, dt,$$

which is bounded by

$$\lim_{n\to\infty} \int_0^1 \int \sup_y |L_n'(y, r_n)|^2 v_n^2(t) \varphi(y) h(t)\, dy\, dt.$$

The proof of (ii) is derived by (i) and the assumption.

Lemma 5.3.4

$$\int_{-r_n}^{r_n} \frac{\{\tilde{f}_n^{(\nu)}(z, r_n) - f^{(\nu)}(z, r_n)\}^2}{f(z, r_n)}\, dz = O_p(n^{-2/3} \log^2 n) r_n^{4\nu+2a_1+6}. \quad (5.3.3)$$

$$\int_{-r_n}^{r_n} \{\tilde{f}_n^{(\nu)}(z, r_n) - f^{(\nu)}(z, r_n)\}^2\, dz = O_p(n^{-2/3} \log^2 n) r_n^{4\nu+2a_1+5}. \quad (5.3.4)$$

Proof. The proofs of (5.3.3) and (5.3.4) are the same. We only prove the first one. The left-hand side of (5.3.3) is bounded by

$$2\int_{-r_n}^{r_n} \Big\{\frac{1}{n}\sum_{j=1}^{k_n} \frac{\psi_{r_n}^{(\nu)}(z - \Lambda_{n2j}, r_n) - f^{(\nu)}(z, r_n)}{f^{1/2}(z, r_n)}\Big\}^2 dz$$

$$+2\int_{-r_n}^{r_n} \Big\{\frac{1}{n}\sum_{j=1+k_n}^{n} \frac{\psi_{r_n}^{(\nu)}(z - \Lambda_{n1j}, r_n) - f^{(\nu)}(z, r_n)}{f^{1/2}(z, r_n)}\Big\}^2 dz$$

$$\stackrel{\text{def}}{=} 2(W_{1n} + W_{2n}).$$

Similarly,

$$W_{1n} \leq 2\int_{-r_n}^{r_n} \Big\{\frac{1}{n}\sum_{j=1}^{k_n} \frac{\psi_{r_n}^{(\nu)}(z - \Lambda_{n2j}, r_n) - \psi_{r_n}^{(\nu)}(z - \Lambda_j, r_n)}{f^{1/2}(z, r_n)}\Big\}^2 dz$$

$$+2\int_{-r_n}^{r_n} \Big\{\frac{1}{n}\sum_{j=1}^{k_n} \frac{\psi_{r_n}^{(\nu)}(z - \Lambda_j, r_n) - f^{(\nu)}(z, r_n)}{f^{1/2}(z, r_n)}\Big\}^2 dz$$

$$\stackrel{\text{def}}{=} (W_{1n}^{(1)} + W_{1n}^{(2)}).$$

By the Taylor expansion,

$$W_{1n}^{(1)} = \int_{-r_n}^{r_n} \Big\{\frac{1}{n}\sum_{j=1}^{k_n} \frac{\psi_{r_n}^{(\nu+1)}(z-\Lambda_j+\mu_{nj}t_{n2j}, r_n)}{f^{1/2}(z, r_n)} t_{n2j}\Big\}^2\, dz,$$

where $t_{n2j} = \Lambda_{n2j} - \Lambda_j$ for $j = 1, \cdots, k_n$ and $\mu_{nj} \in (0,1)$. It follows from the assumptions on $\widehat{\beta}_2$ and $\widehat{g}_2(t)$ that $t_{n2j} = O_p(n^{-1/3}\log n)$. It follows from Lemma 5.3.2 that $W_{1n}^{(1)}$ is smaller than

$$O_p(n^{-2/3}\log^2 n) r_n^{4\nu+2a_1+5} \cdot \int_{-r_n}^{r_n} \frac{1}{n}\sum_{j=1}^{k_n} \frac{\psi_{r_n}(z-\Lambda_j, r_n)}{f(z, r_n)} dz. \tag{5.3.5}$$

Noting that the mean of the latter integrity of (5.3.5) is smaller than $2r_n$, we conclude that

$$W_{1n}^{(1)} = O_p(n^{-2/3}\log^2 n) r_n^{4\nu+2a_1+6}. \tag{5.3.6}$$

Similar to (5.3.5), we have

$$\begin{aligned} EW_{1n}^{(2)} &= \int_{-r_n}^{r_n} \sum_{j=1}^{k_n} \frac{E\{\psi_{r_n}^{(\nu)}(z-\Lambda_j, r_n) - f^{(\nu)}(z, r_n)\}^2}{n^2 f(z, r_n)} dz \\ &\le \int_{-r_n}^{r_n} \sum_{j=1}^{k_n} \frac{E\{\psi_{r_n}^{(\nu)}(z-\Lambda_j, r_n)\}^2}{n^2 f(z, r_n)} dz \\ &\le Cn^{-1} r_n^{4\nu+2a_1+2}. \end{aligned}$$

Thus, we obtain that $W_{1n} = O_p(n^{-2/3}\log^2 n) r_n^{4\nu+2a_1+6}$.

The same conclusion is true for W_{2n}. We therefore complete the proof of Lemma 5.3.4.

5.3.4 Appendix

Proof of Theorem 5.3.1. To prove Theorem 5.3.1 is equivalent to checking Assumptions A.1-A.3 of Schick (1986). The verifications of A.3.1 and A.3.2 are obvious. Lemma 5.3.3 can be applied to finish the verifications of Assumptions A.3.4 and A.3.5 in Schick (1986). We omit the details.

Proof of Theorem 5.3.2. We now provide an outline of the proof of Theorem 5.3.2 as follows. Denote

$$\begin{aligned} A(r_n) &= \int_{-r_n}^{r_n} L_n^2(y, r_n) f(y, r_n)\, dy, \\ \beta_n &= \bar{\bar{\beta}} - \frac{1}{nA(r_n)} \Sigma^{-1} \Big\{\sum_{j=1}^{k_n} X_j L_n(\Lambda_{n2j}, r_n) + \sum_{j=k_n+1}^{n} X_j L_n(\Lambda_{n1j}, r_n)\Big\}. \end{aligned}$$

It is easily shown that $A(r_n) \to I$, which means that β_n is also an **asymptotically efficient** estimator of β when φ is known. If we can show that $\sqrt{n}(\widetilde{\beta}_n - \beta_n)$ converges to zero, then the proof of Theorem 5.3.2 is completed.

Note that

$$\widetilde{\beta}_n - \beta_n = \frac{1}{\widetilde{A}_n(r_n)}(S_{1n} + S_{2n}) + \frac{1}{\widetilde{A}_n(r_n)A(r_n)}(S_{3n} + S_{4n}),$$

where

$$\begin{aligned}
S_{1n} &= \frac{1}{n}\sum_{j=1}^{k_n} X_j\Big\{\int_{-r_n}^{r_n} \widetilde{L}_n(z,r_n)\psi_{r_n}(z-\Lambda_{n2j},r_n)\,dz - L_n(\Lambda_j,r_n)\Big\},\\
S_{2n} &= \frac{1}{n}\sum_{j=k_n+1}^{n} X_j\Big\{\int_{-r_n}^{r_n} \widetilde{L}_n(z,r_n)\psi_{r_n}(z-\Lambda_{n1j},r_n)\,dz - L_n(\Lambda_j,r_n)\Big\},\\
S_{3n} &= \frac{1}{n}\{\widetilde{A}_n(r_n) - A(r_n)\}\sum_{j=1}^{k_n} X_j L_n(\Lambda_{n2j},r_n),\\
S_{4n} &= \frac{1}{n}\{\widetilde{A}_n(r_n) - A(r_n)\}\sum_{j=k_n+1}^{n} X_j L_n(\Lambda_{n1j},r_n).
\end{aligned}$$

To prove Theorem 5.3.2 is equivalent to proving $\|S_{ln}\| = o(n^{-1/2})$ for $l = 1,2,3,4$. We shall complete these proofs in three steps. Step 1 shows $\|S_{1n}\| = o(n^{-1/2})$ and $\|S_{2n}\| = o(n^{-1/2})$. Step 2 analyzes the first part of S_{3n}, $\widetilde{A}_n(r_n) - A(r_n)$, and derives its convergence rate. Step 3 handles the second part of S_{3n}, and then completes the proof of $\|S_{3n}\| = o(n^{-1/2})$. The same arguments may be used for proving $\|S_{4n}\| = o(n^{-1/2})$. We omit the details.

STEP 1 $\|S_{1n}\| = o_p(n^{-1/2})$ and $\|S_{2n}\| = o_p(n^{-1/2})$.

Proof. It is easy to see that

$$\begin{aligned}
S_{1n} &= \int_{-r_n}^{r_n} \{\widetilde{L}_n(z,r_n) - L(z,r_n)\}\Big\{\frac{1}{n}\sum_{j=1}^{k_n} X_j\psi_{r_n}(z-\Lambda_{n2j},r_n)\Big\}\,dz\\
&= \int_{-r_n}^{r_n} \{\widetilde{L}_n(z,r_n) - L(z,r_n)\}\frac{1}{n}\sum_{j=1}^{k_n} X_j\{\psi_{r_n}(z-\Lambda_{n2j},r_n) - f(z,r_n)]\,dz.
\end{aligned}$$

(i) When $n^2 f_n(z,r_n) \le 1$, it follows from Lemma 5.3.1 that

$$\begin{aligned}
\|S_{1n}\| &= \Big\|\int_{-r_n}^{r_n} \{\widetilde{L}_n(z,r_n) - L(z,r_n)\}\frac{1}{n}\sum_{j=1}^{k_n} X_j\psi_{r_n}(z-\Lambda_{n2j},r_n)\,dz\Big\|\\
&\le 2r_n\int_{-r_n}^{r_n}\frac{1}{n}\sum_{j=1}^{k_n}\|X_j\|\psi_{r_n}(z-\Lambda_{n2j},r_n)\,dz\\
&\le 2r_n\int_{-r_n}^{r_n}\frac{1}{n^2}\sum_{j=1}^{k_n}\|X_j\|\,dz\\
&= O_p(n^{-1}r_n^2).
\end{aligned}$$

(ii) When $n^2 f(z, r_n) \le 1$, $\|S_{1n}\|$ is bounded by

$$2r_n \sqrt{\int_{-r_n}^{r_n} \frac{1}{n} \sum_{j=1}^{k_n} \|X_j\|^2} \sqrt{\int_{-r_n}^{r_n} \sum_{j=1}^{k_n} \psi_{r_n}^2(z - \Lambda_{n2j}, r_n)\, dz}.$$

It follows from Lemmas 5.3.2 and 5.3.1 that

$$\begin{aligned}\sum_{j=1}^{k_n} \psi_{r_n}^2(z - \Lambda_{n2j}, r_n) &= O_p(1) \sum_{j=1}^{k_n} \{\psi_{r_n}^2(z - \Lambda_j, r_n) + \psi_{r_n}^{2-2a_1}(z - \Lambda_j, r_n)\} \\ &= O_p(r_n) \sum_{j=1}^{k_n} \psi_{r_n}(z - \Lambda_j, r_n),\end{aligned}$$

and so $\|S_{1n}\| = O_p(r_n^2)\sqrt{\int_{-r_n}^{r_n} f(z, r_n)\, dz} = O_p(n^{-1} r_n^{5/2})$.
(iii) When $n^2 f(z, r_n) > 1$ and $n^2 f_n(z, r_n) > 1$,

$$\begin{aligned}\|S_{1n}\| &\le \int_{-r_n}^{r_n} \left| \frac{\widetilde{f}_n'(z, r_n)}{\widetilde{f}_n(z, r_n)} - \frac{f'(z, r_n)}{f(z, r_n)} \right| \left\| \frac{1}{n} \sum_{j=1}^{k_n} X_j \{\psi_{r_n}(z - \Lambda_{n2j}, r_n) - f(z, r_n)\} \right\| dz \\ &\le \sqrt{\frac{1}{n} \sum_{j=1}^{k_n} \|X_j\|^2} \int_{-r_n}^{r_n} \left| \frac{\widetilde{f}_n'(z, r_n)}{\widetilde{f}_n(z, r_n)} - \frac{f'(z, r_n)}{f(z, r_n)} \right| \\ &\quad \times \sqrt{\left| \frac{1}{n} \sum_{j=1}^{k_n} \{\psi_{r_n}(z - \Lambda_{n2j}, r_n) - f(z, r_n)\}^2 \right|}\, dz. \qquad (5.3.7)\end{aligned}$$

Denote

$$Q_{1n} = \frac{1}{n} \sum_{j=1}^{k_n} \{\psi_{r_n}(z - \Lambda_{n2j}, r_n) - f(z, r_n)\}^2.$$

The integration part of (5.3.7) is bounded by

$$\begin{aligned}&\int_{-r_n}^{r_n} \left| \frac{\widetilde{f}_n'(z, r_n) - f'(z, r_n)}{f(z, r_n)} \right| Q_{1n}^{1/2}\, dz \\ &\quad + \int_{-r_n}^{r_n} \left| \frac{\widetilde{f}_n'(z, r_n)}{\widetilde{f}_n(z, r_n)} \right| \left| \frac{\widetilde{f}_n(z, r_n) - f(z, r_n)}{f(z, r_n)} \right| Q_{1n}^{1/2}\, dz \\ &\le \left\{ \int_{-r_n}^{r_n} \frac{|\widetilde{f}_n'(z, r_n) - f'(z, r_n)|^2}{f(z, r_n)}\, dz \int_{-r_n}^{r_n} \frac{Q_{1n}}{f(z, r_n)}\, dz \right\}^{1/2} \\ &\quad + r_n^{1+a_1} n^{2a_1} \left\{ \int_{-r_n}^{r_n} \frac{|\widetilde{f}_n(z, r_n) - f(z, r_n)|^2}{f(z, r_n)} \int_{-r_n}^{r_n} \frac{Q_{1n}}{f(z, r_n)}\, dz \right\}^{1/2}.\end{aligned}$$

Similar to (5.3.3), we deduce

$$\int_{-r_n}^{r_n} \frac{Q_{1n}}{f(z, r_n)}\, dz = O_p(n^{-2/3} \log^2 n) r_n^{2a_1+6}.$$

This fact and Lemma 5.3.4 deduce

$$\|S_{1n}\| = o_p(n^{-1/2}).$$

A combination of the results in (i), (ii) and (iii) completes the proof of the first part of Step 1. Similarly we can prove the second part. We omit the details.

STEP 2.

$$\widetilde{A}_n(r_n) - A(r_n) = O_p(n^{-2/3+2a_1}\log^2 n) r_n^{3a_1+4}. \tag{5.3.8}$$

Proof. $\widetilde{A}_n(r_n) - A(r_n)$ can be rewritten as

$$\begin{aligned}
\int_{-r_n}^{r_n} \widetilde{L}_n^2(z,r_n)\widetilde{f}_n(z,r_n)\,dz \quad - \quad & \int_{-r_n}^{r_n} L^2(z,r_n)f(z,r_n)\,dz \\
= \quad & \int_{-r_n}^{r_n} \{\widetilde{L}_n^2(z,r_n) - L^2(z,r_n)\}\widetilde{f}_n(z,r_n)\,dz \\
& + \int_{-r_n}^{r_n} L^2(z,r_n)\{\widetilde{f}_n(z,r_n) - f(z,r_n)\}\,dz \\
\stackrel{\text{def}}{=} \quad & I_{1n} + I_{2n}.
\end{aligned}$$

It follows from Lemma 5.3.4 that $I_{2n} = O_p(n^{-2/3}\log^2 n) r_n^{a_1+5}$. We next consider I_{1n}.

(i) When $n^2 f_n(z,r_n) < 1$, $|I_{1n}| \le 2r_n^2 \int_{-r_n}^{r_n} \widetilde{f}_n(z,r_n)\,dz = O(n^{-2}r_n^3)$.

(ii) When $n^2 f(z,r_n) < 1$, it follows from Lemma 5.3.4 that

$$\begin{aligned}
|I_{1n}| \quad & \le 2r_n^2 \Big\{ \sqrt{\int_{-r_n}^{r_n} |\widetilde{f}_n(z,r_n) - f(z,r_n)|^2\,dz r_n^{1/2}} + \int_{-r_n}^{r_n} f(z,r_n)\,dz \Big\} \\
& = O_p(n^{-1/3}\log n) r_n^{a_1+5}.
\end{aligned}$$

(iii) When $n^2 f_n(z,r_n) \ge 1$ and $n^2 f(z,r_n) \ge 1$,

$$\begin{aligned}
|I_{1n}| \quad \le \quad & \int_{-r_n}^{r_n} |\widetilde{L}_n(z,r_n) - L(z,r_n)||\widetilde{L}_n(z,r_n) + L(z,r_n)|\widetilde{f}_n(z,r_n)\,dz \\
\le \quad & 2r_n \int_{-r_n}^{r_n} \left| \frac{\widetilde{f}_n'(z,r_n)}{\widetilde{f}_n(z,r_n)} - \frac{f'(z,r_n)}{f(z,r_n)} \right| \widetilde{f}_n(z,r_n)\,dz \\
\le \quad & 2r_n \Big[\int_{-r_n}^{r_n} |\widetilde{f}_n'(z,r_n) - f'(z,r_n)|\,dz \\
& + \int_{-r_n}^{r_n} \left| \frac{f'(z,r_n)}{f(z,r_n)} \right| |\widetilde{f}_n(z,r_n) - f(z,r_n)|\,dz \Big].
\end{aligned}$$

It follows from Lemma 5.3.4 and `Cauchy-Schwarz inequality` that

$$I_{1n} = O_p(n^{-1/3}\log n) r_n^{a_1+5} + O_p(n^{-1/3+2a_1}\log n) r_n^{3a_1+4}.$$

The conclusions in (i), (ii) and (iii) finish the proof of Step 2.

STEP 3. Obviously, we have that, for $l = 1, \cdots, p$,

$$\begin{aligned}\Big|\sum_{j=1}^{k_n} x_{jl} L_n(\Lambda_{n2j}, r_n)\Big| &= \Big|\int_{-r_n}^{r_n} L(z, r_n) \sum_{j=1}^{k_n} x_{jl} \psi_{r_n}(z - \Lambda_{n2j}, r_n)\, dz\Big| \\ &= \Big|\int_{-r_n}^{r_n} L(z, r_n) \sum_{j=1}^{k_n} x_{jl} \{\psi_{r_n}(z - \Lambda_{n2j}, r_n) - f(z, r_n)\}\, dz\Big|,\end{aligned}$$

which is bounded by

$$\begin{aligned}& r_n \int_{-r_n}^{r_n} \Big|\sum_{j=1}^{k_n} x_{jl} \{\psi_{r_n}(z - \Lambda_{n2j}, r_n) - \psi_{r_n}(z - \Lambda_j, r_n)\}\Big|\, dz \\ & + r_n \int_{-r_n}^{r_n} \Big|\sum_{j=1}^{k_n} x_{jl} \{\psi_{r_n}(z - \Lambda_j, r_n) - f(z, r_n)\}\Big|\, dz \\ & \stackrel{\text{def}}{=} Q_{2n}^{(l)} + Q_{3n}^{(l)}.\end{aligned}$$

$Q_{2n}^{(l)}$ is bounded by

$$r_n \int_{-r_n}^{r_n} \sqrt{\sum_{j=1}^{k_n} x_{jl}^2} \sqrt{\sum_{j=1}^{k_n} \{\psi_{r_n}(z - \Lambda_{n2j}, r_n) - \psi_{r_n}(z - \Lambda_j, r_n)\}^2\, dz}.$$

The term in the second integration part is just $n \int_{-r_n}^{r_n} Q_{1n}\, dz$. We conclude from the discussion for Q_{1n} that $Q_{2n}^{(l)} = O_p(n^{2/3} \log n) r_n^{a_1 + 4}$.

We now analyze the term $Q_{3n}^{(l)}$. The same way as for $Q_{2n}^{(l)}$ leads that

$$Q_{3n}^{(l)} \le r_n^{3/2} \Big\{\int_{-r_n}^{r_n} \Big|\sum_{j=1}^{k_n} x_{jl} \{\psi_{r_n}(z - \Lambda_j, r_n) - f(z, r_n)\}\Big|^2 dz\Big\}^{1/2}.$$

Note that the expectation value of the integration equals

$$\int_{-r_n}^{r_n} \sum_{j=1}^{k_n} E\, |x_{jl} \{\psi_{r_n}(z - \Lambda_j, r_n) - f(z, r_n)\}|^2\, dz,$$

which is bounded by

$$\begin{aligned}\int_{-r_n}^{r_n} \sum_{j=1}^{k_n} E\, |x_{jl} \psi_{r_n}(z - \Lambda_j, r_n)|^2\, dz &\le \int_{-r_n}^{r_n} \sum_{j=1}^{k_n} E\, |x_{jl}|^2 r_n f(z, r_n)\, dz \\ &= O(n r_n).\end{aligned}$$

We thus conclude that

$$\Big\|\frac{1}{n} \sum_{j=1}^{k_n} X_j L_n(\Lambda_{n2j}, r_n)\Big\| = O_p(n^{-1/3 + 2a_1} \log n) r_n^{a_1 + 4}. \tag{5.3.9}$$

A combination of (5.3.8) and (5.3.9) implies

$$\|S_{3n}\| = o_p(n^{-1/2}).$$

We finally complete the proof of Theorem 5.3.2.

5.4 Bahadur Asymptotic Efficiency

5.4.1 Definition

This section is concerned with large deviations of estimation. Bahadur asymptotic efficiency, which measures the rate of **tail probability**, is considered here. It can be stated (under certain regularity conditions) that, for any consistent estimator $T_n(\mathbf{Q})$,

$$\liminf_{\zeta\to 0}\liminf_{n\to\infty}\frac{1}{n\zeta^2}\log P_\beta\{|T_n(\mathbf{Q})-\beta|\geq\zeta\}\geq -I(\beta)/2,$$

and that the **maximum likelihood estimator** (*MLE*) β_n can achieve the lower bound, that is,

$$\lim_{\zeta\to 0}\lim_{n\to\infty}\frac{1}{n\zeta^2}\log P_\beta\{|\beta_n-\beta|\geq\zeta\}=-I(\beta)/2,$$

where $I(\beta)$ is Fisher's information. In other words, for any consistent estimator T_n, $P_\beta\{|T_n(\mathbf{Q})-\beta|\geq\zeta\}$ cannot tend to zero faster than the exponential rate given by $\exp\{-n/2\zeta^2 I(\beta)\}$, and for *MLE* β_n, $P_\beta\{|\beta_n-\beta|\geq\zeta\}$ achieves this **optimal exponential rate**. The β_n is called **Bahadur asymptotically efficient** (**BAE**).

Fu (1973) showed, under regularity conditions which differ partly from Bahadur's, that a large class of consistent estimators β_n^* are asymptotically efficient in Bahadur's sense. The author also gave a simple and direct method to verify Bahadur (1967) conditions. Cheng (1980) proved, under weaker conditions than Bahadur's, that the *MLE* in both single-parameter and multi-parameter cases is **BAE**. Lu (1983) studied the Bahadur efficiency of *MLE* for the linear models.

In this section, we investigate BAE of the estimator of the parameter β in the model (1.1.1). It is shown that a **quasi-maximum likelihood** estimator (QMLE) is BAE under suitable assumptions. Similar results for generalized semiparametric models have been established by Liang (1995a).

In Sections 5.4 and 5.5 below, we always suppose that g is an unknown **Hölder continuous** function of known order of $m+r$ (Chen, 1988) in R^1. When approximating g, we use a **piecewise polynomial** approximation $\widehat{g}_P$ studied by Stone (1985) and Chen (1988), which satisfies

$$|g(T_i)-\widehat{g}_P(T_i)|\leq B_2 M_n^{-(m+r)}\quad i=1,\cdots,n, \tag{5.4.1}$$

where M_n satisfies $\lim_{n\to\infty} nM_n^{-2(m+r)} = 0$ and $\lim_{n\to\infty} n^{-q}M_n = 0$ for some $q \in (0,1)$.

The *MLE* $\widehat{\beta}_{ML}$ of β is defined based on $\{Y_i = X_i^T\beta + \widehat{g}_P(T_i) + \varepsilon_i \quad i = 1, \cdots, n\}$.

Let $\{X_i, T_i, Y_i, i = 1, \cdots, n\}$ be a sequence of i.i.d. observations drawn from the model (1.1.1). Throughout the remainder of this section we denote a vector by a boldface letter, a matrix by a calligraphic letter, $\mathcal{R}_n^* = \mathbf{X}^T\mathbf{X}$, $I = I(\varphi) = \int (\psi'(x))^2 \varphi(x)dx < \infty$, and $\psi(x) = \varphi'(x)/\varphi(x)$. We assume that $\varphi(x)$ is positive and twice differentiable in R^1 and that the limit values of $\varphi(x)$ and $\varphi'(x)$ are zero as $x \to \infty$. For $a \in R^p$ and $\mathcal{B}_1 = (b_{ij})_{n_1 \times n_2}$, denote

$$\|a\| = \Big(\sum_{i=1}^{p} a_i^2\Big)^{1/2}, \quad |a| = \max_{1\le i\le p} |a_i|;$$
$$|\mathcal{B}_1|^* = \max_{i,j} |b_{ij}|, \quad \|\mathcal{B}_1\|^* = \max_{a\in R^{n_2}, \|a\|=1} \|\mathcal{B}_1 a\|,$$

where $\|\cdot\|$ denotes L_2-norm and $\|\cdot\|^*$ does matrix norm.

We now state the following definitions.

Definition 5.4.1 *The estimator $\widetilde{h}_n(Y_1, \cdots, Y_n)$ of β is called* `locally uniformly consistent estimator` *of β, if for every $\beta_0 \in R^p$, there exists a $\delta > 0$ such that for each $\zeta > 0$,*

$$\lim_{n\to\infty} \sup_{|\beta-\beta_0|<\delta} P_\beta\{\|\widetilde{h}_n - \beta\| > \zeta\} = 0.$$

Definition 5.4.2 *Assume that ${\mathcal{R}_n^*}^{-1}$ exists. The consistent estimator $\widetilde{h}_n$ of β is said to be* `BAE`*, if for each $\beta_0 \in R^p$,*

$$\limsup_{\zeta\to 0} \limsup_{n\to\infty} \frac{1}{\zeta^2} \|{\mathcal{R}_n^*}^{-1}\|^* \log P_{\beta_0}\{\|\widetilde{h}_n - \beta_0\| > \zeta\} \le -\frac{I}{2}.$$

5.4.2 Tail Probability

We make the following assumptions for our main result.

Assumption 5.4.1 *There exist constants C_1, $C_2 > 0$ such that $C_1 \le \mu_1 \le \mu_2 \le \cdots \le \mu_p \le C_2$, where $\mu_1, \mu_2, \cdots, \mu_p$ are the eigenvalues of $n^{-1}\mathcal{R}_n^*$. Denote $R = C_2/C_1$.*

Assumption 5.4.2 *There exists a C_0 such that $E\|X\| \le C_0 < \infty$.*

Assumption 5.4.3 $\lim_{\delta\to 0} \int \sup_{|h|\le\delta} |\psi'(y+h) - \psi'(y)|\varphi(y)\, dy = 0.$

Assumption 5.4.4 *There exists a* $t_0 > 0$ *such that*

$$\int \exp\{t_0|\psi(x)|\}\varphi(x)\,dx < \infty \text{ and } \int \exp\{t_0|\psi'(x)|\}\varphi(x)\,dx < \infty.$$

Assumption 5.4.5 *There exist a measurable function* $h(x) > 0$ *and a nondecreasing function* $\gamma(t)$ *satisfying* $\gamma(t) > 0$ *for* $t > 0$ *and* $\lim_{t\to 0^+}\gamma(t) = 0$ *such that* $\int \exp\{h(x)\}\varphi(x)\,dx < \infty$ *and* $|\psi'(x+t) - \psi'(x)| \le h(x)\gamma(t)$ *whenever* $|t| \le |t_0|$.

Assumption 5.4.6 *The MLE* $\widehat{\beta}_{ML}$ *exists, and for each* $\delta > 0$, $\beta_0 \in R^p$, *there exist constants* $K = K(\delta, \beta_0)$ *and* $\rho = \rho(\delta, \beta_0) > 0$ *such that*

$$P_{\beta_0}\Big\{|\widehat{\beta}_{ML} - \beta_0| > \delta\Big\} \le K\exp(-\rho\|\mathcal{R}_n^*\|^*\delta^2).$$

The following theorem gives our main result which shows that $\widehat{\beta}_{ML}$ is a *BAE* estimator of β.

Theorem 5.4.1 $\widetilde{h}_n$ *is a* `locally uniformly consistent estimator`. *Suppose that Assumptions 5.4.1-5.4.3 hold. Then for each* $\beta_0 \in R^p$

$$\liminf_{\zeta\to 0}\liminf_{n\to\infty}\frac{1}{\zeta^2}\|\mathcal{R}_n^{*-1}\|^*\log P_{\beta_0}\{\|\widetilde{h}_n - \beta_0\| > \zeta\} \ge -\frac{I}{2}. \tag{5.4.2}$$

If Assumptions 5.4.1-5.4.6 hold, then for each $\beta_0 \in R^p$

$$\limsup_{\zeta\to 0}\limsup_{n\to\infty}\frac{\|\mathcal{R}_n^{*-1}\|^*}{\zeta^2}\log P_{\beta_0}\Big\{\|\widehat{\beta}_{ML} - \beta_0\| > \zeta\Big\} \le -\frac{I}{2}. \tag{5.4.3}$$

The result (5.4.3) implies that $\widehat{\beta}_{ML}$ is BAE.

5.4.3 Technical Details

The proof of the theorem is partly based on the following lemma, which was proven by Lu (1983).

Lemma 5.4.1 *Assume that* $W_1, \cdots, W_n$ *are i.i.d. with mean zero and finite variance* σ_1^2. *There exists a* $t_0 > 0$ *such that* $E\{\exp(t_0|W_1|)\} < \infty$. *For known constants* $a_{1n}, a_{2n}, \cdots, a_{nn}$, *there exist constants* A_n *and* $\widehat{A}$ *such that* $\sum_{i=1}^n a_{in}^2 \le A_n$ *and* $\max_{1\le i\le n}|a_{in}|/A_n \le \widehat{A}$. *Then for small enough* $\zeta > 0$,

$$P\Big\{\Big|\sum_{i=1}^n a_{in}W_i\Big| > \zeta\Big\} \le 2\exp\Big\{-\frac{\zeta^2}{2{\sigma_1}^2A_n^2}(1+o(\zeta))\Big\},$$

where $|o(\zeta)| \le \widehat{A}C_1\zeta$ *and* C_1 *only depends on* W_1.

We first prove the result (5.4.2). The proof is divided into three steps. In the first step, we get a uniformly most powerful (UMP) test Φ_n^* whose power is 1/2 for the hypothesis $H_0 : \beta = \beta_0 \Longleftrightarrow H_1 : \beta = \beta_n$. In the second step, by constructing a test $\Phi_n(\mathbf{Y})$ corresponding to $\widetilde{h}_n$, we show that the power of the constructed test is larger than 1/2. In the last step, by using `Neyman-Pearson Lemma`, we show that $E_{\beta_0}\Phi_n$, the level of Φ_n, is larger than $E_{\beta_0}\Phi_n^*$.

Proof of (5.4.2). For each $\zeta > 0$, set

$$\beta_n = \beta_0 + \frac{\mathcal{R}_n^{*-1} a_n}{\|\mathcal{R}_n^{*-1}\|^*}\zeta,$$

where $a_n \in R^p$, $a_n^T \mathcal{R}_n^{*-1} a_n = \|\mathcal{R}_n^{*-1}\|^*$ and $\|a_n\| = 1$. Let $l_i = (0, \cdots, 1, \cdots, 0)^T \in R^p$, it is easy to get

$$\|\mathcal{R}_n^{*-1}\|^* \geq a^T \mathcal{R}_n^{*-1} a \geq \frac{1}{\|\mathcal{R}_n^*\|^*}$$

and

$$|\beta_n - \beta_0| \leq \|\mathcal{R}_n^*\| \cdot \|\mathcal{R}_n^{*-1}\|^* \zeta \leq R\zeta.$$

Denote

$$\begin{aligned} \Gamma_n(\mathbf{Y}) &= \prod_{i=1}^n \frac{\varphi(Y_i - X_i^T\beta_n - g(T_i))}{\varphi(Y_i - X_i^T\beta_0 - g(T_i))}, \\ \Delta_i &= X_i^T(\beta_n - \beta_0), \quad d_n = \exp\left\{\frac{I(1+\mu)\zeta^2}{2\|\mathcal{R}_n^{*-1}\|^*}\right\} \quad (\mu > 0). \end{aligned}$$

By `Neyman-Pearson Lemma`, there exists a test $\Phi_n^*(\mathbf{Y})$ such that

$$E_{\beta_n}\{\Phi_n^*(\mathbf{Y})\} = \frac{1}{2}.$$

Under H_0, we have the following inequality:

$$\begin{aligned} E_{\beta_0}\{\Phi_n^*(\mathbf{Y})\} &\geq \int_{\Gamma_n \leq d_n} \Phi_n^*(\mathbf{Y})\, dP_{n\beta_0} \\ &\geq \frac{1}{d_n}\Big\{\frac{1}{2} - \int_{\Gamma_n(\mathbf{Y}) > d_n} \Phi_n^*(\mathbf{Y})\, dP_{n\beta_n}\Big\}. \end{aligned}$$

If

$$\limsup_{n\to\infty} P_{\beta_n}\{\Gamma_n(\mathbf{Y}) > d_n\} \leq \frac{1}{4}, \tag{5.4.4}$$

then for n large enough

$$E_{\beta_0}\{\Phi_n^*(\mathbf{Y})\} \geq \frac{1}{4d_n}. \tag{5.4.5}$$

Define

$$\Phi_n(Y) = \begin{cases} 1, & |a_n^T(\widetilde{h}_n - \beta_0)| \geq \lambda'\zeta \\ 0, & \text{otherwise,} \end{cases} \tag{5.4.6}$$

where $\lambda' \in (0,1)$. Since $a_n^T(\beta_n - \beta_0) = \zeta$ and $\widetilde{h}_n$ is a `locally uniformly consistent estimator`, we have

$$\liminf_{n\to\infty} E_{\beta_n}\{\Phi_n(\mathbf{Y})\} \geq \liminf_{n\to\infty} P_{\beta_n}\{\|\widetilde{h}_n - \beta_n\| \leq (1-\lambda')\zeta\} = 1.$$

It follows from `Neyman-Pearson Lemma` that for n large enough

$$E_{\beta_0}\{\Phi_n(\mathbf{Y})\} \geq E_{\beta_0}\{\Phi_n^*(\mathbf{Y})\}. \tag{5.4.7}$$

It follows from (5.4.5), (5.4.6) and (5.4.7) that

$$\begin{aligned} P_{\beta_0}\{\|\widetilde{h}_n - \beta_0\| \geq \lambda'\zeta\} &\geq P_{\beta_0}\{|a_n^T(\widetilde{h}_n - \beta_0)| \geq \lambda'\zeta\} \\ &= E_{\beta_0}\{\Phi_n(\mathbf{Y})\} \geq \frac{1}{4d_n}. \end{aligned}$$

By letting $\mu \to 0$ and $\lambda' \to 1$, this completes the proof of (5.4.2).

Now we return to prove the inequality (5.4.4). It follows from Assumption 5.4.2 that $\|\Delta_i\| \leq RC\zeta$ and

$$\sum_{i=1}^n \Delta_i{}^2 \leq \frac{2a_n^T \mathcal{R}_n^{*\,-1} a_n}{\|\mathcal{R}_n^{*\,-1}\|^{*2}}\zeta^2 = \frac{2\zeta^2}{\|\mathcal{R}_n^{*\,-1}\|^*}. \tag{5.4.8}$$

A Taylor expansion implies that for sufficiently small ζ

$$\sum_{i=1}^n \log\frac{\varphi(Y_i)}{\varphi(Y_i+\Delta_i)} = -\sum_{i=1}^n\Big\{\psi(Y_i)\Delta_i + \frac{1}{2}(\psi'(Y_i) + R_i(Y_i))\Delta_i^2\Big\},$$

where $R_i(Y_i) = \psi'(Y_i + \theta_i\Delta_i) - \psi'(Y_i)$ and $0 < \theta_i < 1$. Thus

$$\begin{aligned} P_{\beta_n}\{\Gamma_n(Y) > d_n\} &= P_0\left\{\prod_{i=1}^n \frac{\varphi(Y_i)}{\varphi(Y_i+\Delta_i)} > d_n\right\} \\ &= P_0\left\{\sum_{i=1}^n \log\frac{\varphi(Y_i)}{\varphi(Y_i+\Delta_i)} > \frac{I(1+\mu)\zeta^2}{2\|\mathcal{R}_n^{*\,-1}\|^*}\right\} \\ &\leq P_0\left\{\frac{1}{2}\sum_{i=1}^n I(\varphi)\Delta_i^2 > \frac{I(1+\frac{\mu}{2})\zeta^2}{2\|\mathcal{R}_n^{*\,-1}\|^*}\right\} \\ &\quad + P_0\left\{-\sum_{i=1}^n \psi(Y_i)\Delta_i > \frac{I\mu\zeta^2}{12\|\mathcal{R}_n^{*\,-1}\|^*}\right\} \\ &\quad + P_0\left\{-\sum_{i=1}^n \frac{1}{2}[\psi'(Y_i) + I(\varphi)]\Delta_i^2 > \frac{I\mu\zeta^2}{12\|\mathcal{R}_n^{*\,-1}\|^*}\right\} \\ &\quad + P_0\left\{-\frac{1}{2}\sum_{i=1}^n R_i(Y_i)\Delta_i^2 > \frac{I\mu\zeta^2}{12\|\mathcal{R}_n^{*\,-1}\|^*}\right\} \\ &\stackrel{\text{def}}{=} P_1 + P_2 + P_3 + P_4. \end{aligned}$$

We now estimate the four probabilities P_1, P_2, P_3 and P_4, respectively.

$$P_1 = P_0\left\{ I(\varphi)\frac{1}{\|\mathcal{R}_n^{*-1}\|^*} > \frac{I(1+\frac{\mu}{2})}{\|\mathcal{R}_n^{*-1}\|^*}\right\} = 0. \tag{5.4.9}$$

It follows from **Chebyshev's inequality**, Assumption 5.4.1 and (5.4.8) that

$$\begin{aligned} P_2 &= P_{\mathbf{T}}\Big[P_0\Big\{-\sum_{i=1}^{n}\psi(Y_i)\Delta_i > \frac{I\mu\zeta^2}{12\|\mathcal{R}_n^{*-1}\|^*}\Big|\mathbf{T}\Big\}\Big] \\ &\le P_{\mathbf{T}}\Big(\frac{144\|\mathcal{R}_n^{*-1}\|^*}{I\mu^2\zeta^2}\Big) \to 0 \quad \text{as } n\to\infty. \end{aligned} \tag{5.4.10}$$

$$P_3 \le P_{\mathbf{T}}\Big[\Big(\frac{6\|\mathcal{R}_n^{*-1}\|^*}{I\mu\zeta^2}\Big)^2 (CR\zeta)^2 \frac{\zeta^2}{\|\mathcal{R}_n^{*-1}\|^*} E_0\{\psi'(Y_1)+I(\varphi)\}^2\Big] \to 0 \quad \text{as } n\to\infty.$$

$$P_4 \le P_{\mathbf{T}}\Big[\frac{6\|\mathcal{R}_n^{*-1}\|^*}{I\mu\zeta^2}\cdot\frac{1}{\|\mathcal{R}_n^{*-1}\|^*}\zeta^2 E_0\{\max_{1\le i\le n}|R_i(Y_i)|\Big|\mathbf{T}\}\Big].$$

By Assumption 5.4.3 and letting ζ be sufficiently small such that $|\Delta_i| < \delta$, we have

$$E_0\{\max_{1\le i\le n}|R_i(Y_i)|\Big|\mathbf{T}\} \le \int \sup_{|h|<\delta}|\psi'(y+h)-\psi'(y)|\varphi(y)\,dy \le \frac{\mu}{24}. \tag{5.4.11}$$

Combining the results (5.4.9) to (5.4.11), we finally complete the proof of (5.4.4).

We outline the proof of the formula (5.4.3). Firstly, by using the Taylor expansion and Assumption 5.4.2, we get the expansion of the projection, $a^T(\widehat{\beta}_{ML}-\beta)$, of $\widehat{\beta}_{ML}-\beta$ on the unit sphere $\|a\|=1$. Secondly we decompose $P_{\beta_0}\{|a^T(\widehat{\beta}_{ML}-\beta_0)| > \zeta\}$ into five terms. Finally, we calculate the value of each term.

It follows from the definition of $\widehat{\beta}_{ML}$ and Assumption 5.4.2 that

$$\begin{aligned} a^T(\widehat{\beta}_{ML}-\beta) &= -\{(\mathcal{F}I)^{-1}+\widetilde{W}\}\sum_{i=1}^{n}\psi(Y_i - X_i^T\beta - g(T_i))a^T\mathcal{R}_n^{*-1} \\ &\quad -\sum_{i=1}^{n}\psi'(Y_i - X_i^T\beta^* - g^*(T_i))a^T\mathcal{R}_n^{*-1}X_i\{\widehat{g}_P(T_i)-g(T_i)\}, \end{aligned}$$

where $X_i^T\beta^*$ lies between $X_i^T\widehat{\beta}_{ML}$ and $X_i^T\beta$ and $g^*(T_i)$ lies between $g(T_i)$ and $\widehat{g}_P(T_i)$.

Denote

$$\begin{aligned} &R_i(Y_i,X_i,T_i) = \psi'(Y_i - X_i^T\beta^* - g^*(T_i)) - \psi'(Y_i - X_i^T\beta - g(T_i)), \\ &R_1^* = \sum_{i=1}^{n}\{I+\psi'(Y_i - X_i^T\beta - g(T_i))\}\mathcal{R}_n^{*-1}X_iX_i^T, \\ &R_2^* = \sum_{i=1}^{n}R_i(Y_i,X_i,T_i)\mathcal{R}_n^{*-1}X_iX_i^T. \end{aligned}$$

Let α be sufficiently small such that $\det(-\mathcal{F}I+R_1^*+R_2^*)\neq 0$ when $|R_1^*+R_2^*|<\alpha$. Hence $(-\mathcal{F}I+R_1^*+R_2^*)^{-1}$ exists and we denote it by $-\{(\mathcal{F}I)^{-1}+\widetilde{W}\}$. Moreover, according to continuity, there is a nondecreasing and positive function $\eta(\alpha)$ such that $|\widetilde{W}|<\eta(\alpha)$ and $\lim_{\alpha\to 0}\eta(\alpha)=0$ when $|R_1^*+R_2^*|<\alpha$. Hence, we obtain, for every $0<\lambda<1/4$, $a\in R^p$ and $\|a\|=1$, that $P_{\beta_0}\{|a^T(\widehat{\beta}_{ML}-\beta_0)|>\zeta\}$ is bounded by

$$\begin{aligned}
& P_{\beta_0}\Big\{\Big|\sum_{i=1}^n \psi(Y_i-X_i^T\beta_0-g(T_i))a^T\mathcal{R}_n^{*-1}X_i\Big|>(1-2\lambda)I\zeta\Big\}\\
& +P_{\beta_0}\Big\{\Big|\sum_{i=1}^n a^T\psi(Y_i-X_i^T\beta_0-g(T_i))\mathcal{R}_n^{*-1}X_i\Big|>\frac{\lambda\zeta}{\eta(2\alpha)}\Big\}\\
& +P_{\beta_0}\Big\{|R_1^*|>\alpha\Big\}+P_{\beta_0}\{|R_2^*|>\alpha\}\\
& +P_{\beta_0}\Big\{\sum_{i=1}^n|\widehat{g}_P(T_i)-g(T_i)||\psi'(Y_i-X_i^T\beta^*-g^*(T_i))a^T\mathcal{R}_n^{*-1}X_i|>\lambda\zeta\Big\}\\
\stackrel{\text{def}}{=}\ & P_5+P_6+P_7+P_8+P_9.
\end{aligned}$$

In the following we use Lemma 5.4.1 to calculate $\{P_i; 5\le i\le 9\}$. We calculate only the probability P_8 and the others can be obtained similarly. We omit the details.

It follows from Assumption 5.4.5 that

$$\begin{aligned}
|R_i(Y_i,X_i,T_i)| &= |\psi'(Y_i-X_i^T\beta^*-g^*(T_i)-\psi'(Y_i-X_i^T\beta_0-g(T_i))|\\
&\le h(Y_i-X_i^T\beta_0-g(T_i))\gamma(X_i^T(\widehat{\beta}_{ML}-\beta_0)-(g^*(T_i)-g(T_i))).
\end{aligned}$$

Denote $h_0=E_{\beta_0}h(Y_i-X_i^T\beta_0-g(T_i))$. We now have

$$\begin{aligned}
P_8 &\le \sum_{j=1}^p\sum_{s=1}^p P_{\beta_0}\Big\{\Big|\sum_{i=1}^n h(Y_i-X_i^T\beta_0-g(T_i))\gamma(X_i^T(\widehat{\beta}_{ML}-\beta_0)\\
&\qquad -(g^*(T_i)-g(T_i)))l_j^T\mathcal{R}_n^{*-1}X_iX_i^Tl_s\Big|\ge\alpha\Big\}\\
&\le \sum_{j=1}^p\sum_{s=1}^p\Big[P_{\beta_0}\{\cup_{i=1}^n|\widehat{g}_P(T_i)-g(T_i)|>\delta\}+P_{\beta_0}\{|\widehat{\beta}_{ML}-\beta_0|\ge\delta\}\\
&\qquad +P_{\beta_0}\Big\{\Big|\sum_{i=1}^n h(Y_i-X_i^T\beta_0-g(T_i))\gamma(C\delta+\delta)l_j^T\mathcal{R}_n^{*-1}X_iX_i^Tl_s\Big|\ge\alpha\Big\}\Big]\\
&\stackrel{\text{def}}{=} P_8^{(1)}+P_8^{(2)}+P_8^{(3)}.
\end{aligned}$$

Let $\zeta\le\Big(\dfrac{2\rho}{(1-2\lambda)^2I}\Big)^{1/2}$. It follows from Assumption 5.4.6 and $|a^T\mathcal{R}_n^{*-1}a|\ge(\|\mathcal{R}_n^*\|^*)^{-1}$ that

$$\rho(\delta,\beta_0)\|\mathcal{R}_n^*\|^*\ge\frac{(1-2\lambda)^2I}{2a^T\mathcal{R}_n^{*-1}a}\zeta^2$$

and

$$P_8^{(2)} \le K(\delta, \beta_0) \exp\Big\{-\frac{(1-2\lambda)^2 I}{2a^T \mathcal{R}_n^{*-1} a}\zeta^2\Big\}. \tag{5.4.12}$$

Let $\sigma_h^2 = E_{\beta_0}\{h(Y_i - X_i^T\beta_0 - g(T_i)) - h_0\}^2$. It follows from Lemma 5.4.1 that

$$\begin{aligned} P_8^{(3)} &\le P_{\beta_0}\Big[\Big|\sum_{i=1}^n \{h(Y_i - X_i^T\beta_0 - g(T_i)) - h_0\} l_j^T \mathcal{R}_n^{*-1} X_i X_i^T l_s\Big| \ge \frac{\alpha}{2\gamma(C\delta+\delta)}\Big] \\ &\le 2\exp\left\{-\frac{\alpha^2}{8\gamma^2(C\delta+\delta)C^2R\sigma_h^2 a^T\mathcal{R}_n^{*-1}a}\Big(1+O_4\big(\frac{\alpha}{2\gamma(C\delta+\delta)}\big)\Big)\right\}, \end{aligned}$$

where $\Big|O_4\big(\frac{\alpha}{2\gamma(C\delta+\delta)}\big)\Big| \le B_4\frac{\alpha}{2\gamma(C\delta+\delta)}$ and B_4 depends only on $h(\varepsilon_1)$. Furthermore, for ζ small enough, we obtain

$$P_8^{(3)} \le 2\exp\Big\{-\frac{(1-2\lambda)^2 I\zeta^2}{2a^T\mathcal{R}_n^{*-1}a}\Big\}.$$

It follows from (5.4.1) that

$$P_8^{(1)} = P_{\beta_0}\{\cup_{i=1}^n |(\widehat{g}_P(T_i) - g(T_i))| > \delta\} = 0. \tag{5.4.13}$$

Combining (5.4.12) with (5.4.13), we have

$$P_8 \le (2+K)\exp\Big\{-\frac{(1-2\lambda)^2 I\zeta^2}{2a^T\mathcal{R}_n^{*-1}a}\Big\}.$$

Therefore

$$\begin{aligned} P_{\beta_0}\Big\{|a^T(\widehat{\beta}_{ML} - \beta_0)| > \zeta\Big\} \le (Kp^2 &+ K + 4p^2 + 2p) \\ &\times\exp\Big\{-\frac{(1-2\lambda)^2 I\zeta^2}{2a^T\mathcal{R}_n^{*-1}a}(1+O(\zeta))\Big\}, \end{aligned}$$

where $O(\zeta) = \min\{O_1(\zeta), O_2(\zeta), O_3(\zeta), O_4(\zeta), O_5(\zeta)\}$, $|O(\zeta)| \le CRB(\eta^{-1}(2\alpha) + 1)$, which implies

$$\limsup_{\zeta\to 0}\limsup_{n\to\infty} \frac{a^T\mathcal{R}_n^{*-1}a}{\zeta^2}\log P_{\beta_0}\Big\{|a^T(\widehat{\beta}_{ML} - \beta_0)| > \zeta\Big\} \le -\frac{(1-2\lambda)^2 I}{2}.$$

Since a is arbitrary, the result (5.4.3) follows from $\lambda \to 0$. This completes the proof of (5.4.3).

5.5 Second Order Asymptotic Efficiency

5.5.1 Asymptotic Efficiency

In Section 5.3, we constructed a class of asymptotically efficient estimators of the parameter β, that is, their asymptotic variances reach the asymptotically efficient bound, the inverse of the Fisher information matrix. There is plenty of evidence that there exist many asymptotic efficient estimators. The comparison of their strength and weakness is an interesting issue in both theoretical and practical aspects (Liang, 1995b, Linton, 1995). This section introduces a basic result for the partially linear model (1.1.1) with $p = 1$. The basic results are related to the second order asymptotic efficiency. The context of this section is influenced by the idea proposed by Akahira and Takeuchi (1981) for the linear model.

This section consider only the case where ε_i are i.i.d. and (X_i, T_i) in model (1.1.1) are random design points. Assume that X is a covariate variable with finite second moment. For easy reference, we introduce some basic concepts on asymptotic efficiency. We refer the details to Akahira and Takeuchi (1981).

Suppose that Θ is an open set of R^1. A $\{C_n\}$-consistent estimator β_n is called second order asymptotically median unbiased (or second order AMU) estimator if for any $v \in \Theta$, there exists a positive number δ such that

$$\limsup_{n\to\infty} \sup_{\beta:|\beta-v|<\delta} C_n \Big| P_{\beta,n}\{\beta_n \le \beta\} - \frac{1}{2}\Big| = 0$$

and

$$\limsup_{n\to\infty} \sup_{\beta:|\beta-v|<\delta} C_n \Big| P_{\beta,n}\{\beta_n \ge \beta\} - \frac{1}{2}\Big| = 0.$$

Suppose that β_n is a second order AMU estimator. $G_0(t,\beta) + C_n^{-1} G_1(t,\beta)$ is called the second order asymptotic distribution of $C_n(\beta_n - \beta)$ if

$$\lim_{n\to\infty} C_n \Big| P_{\beta,n}\{C_n(\beta_n - \beta) \le t\} - G_0(t,\beta) - C_n^{-1} G_1(t,\beta)\Big| = 0.$$

Let $C_n = \sqrt{n}$ and $\beta_0 (\in \Theta)$ be arbitrary but fixed. We consider the problem of testing hypothesis

$$H^+ : \beta = \beta_1 = \beta_0 + \frac{u}{\sqrt{n}} (u > 0) \longleftrightarrow K : \beta = \beta_0.$$

Denote $\Phi_{1/2} = \Big\{\phi_n : E_{\beta_0 + u/\sqrt{n}, n} \phi_n = 1/2 + o(1/\sqrt{n})\Big\}$ and $A_{\beta_n,\beta_0} = \{\sqrt{n}(\beta_n - \beta_0) \le u\}$. It follows that

$$\lim_{n\to\infty} P_{\beta_0 + \frac{u}{\sqrt{n}}, n}(A_{\beta_n,\beta_0}) = \lim_{n\to\infty} P_{\beta_0 + \frac{u}{\sqrt{n}}, n}\Big\{\beta_n \le \beta_0 + \frac{u}{\sqrt{n}}\Big\} = \frac{1}{2}.$$

Obviously, the indicator functions $\{\mathcal{X}_{A_{\beta_n,\beta_0}}\}$ of the sets $A_{\beta_n,\beta_0}(n=1,2,\cdots)$ belong to $\Phi_{1/2}$. By Neyman-Pearson Lemma, if

$$\sup_{\phi_n\in\Phi_{1/2}} \limsup_{n\to\infty} \sqrt{n}\Big\{E_{\beta_0,n}(\phi_n) - H_0^+(t,\beta_0) - \frac{1}{\sqrt{n}}H_1^+(t,\beta_0)\Big\} = 0,$$

then $G_0(t,\beta_0) \leq H_0^+(t,\beta_0)$. Furthermore, if $G_0(t,\beta_0) = H_0^+(t,\beta_0)$, then $G_1(t,\beta_0) \leq H_1^+(t,\beta_0)$.

Similarly, we consider the problem of testing hypothesis

$$H^- : \beta = \beta_0 + \frac{u}{\sqrt{n}}(u<0) \longleftrightarrow K : \beta = \beta_0.$$

We conclude that if

$$\inf_{\phi_n\in\Phi_{1/2}} \liminf_{n\to\infty} \sqrt{n}\Big\{E_{\beta_0,n}(\phi_n) - H_0^-(t,\beta_0) - \frac{1}{\sqrt{n}}H_1^-(t,\beta_0)\Big\} = 0,$$

then $G_0(t,\beta_0) \leq H_0^-(t,\beta_0)$; if $G_0(t,\beta_0) = H_0^-(t,\beta_0)$, then $G_1(t,\beta_0) \leq H_1^-(t,\beta_0)$.

These arguments indicate that even the second order asymptotic distribution of `second AMU estimator` cannot certainly reach the asymptotic distribution bound. We first introduce the following definition. Its detailed discussions can be found in Akahira and Takeuchi (1981).

Definition 5.5.1 *β_n is said to be* `second order asymptotically efficient` *if its second order asymptotic distribution uniformly attains the bound of the second order asymptotic distribution of* `second order AMU` *estimators, that is for each $\beta \in \Theta$*

$$G_i(u,\beta) = \begin{cases} H_i^+(u,\beta) & \text{for } u>0 \\ H_i^-(u,\beta) & \text{for } u<0. \end{cases}$$

The goal of this section is to consider second order asymptotic efficiency for estimator of β.

5.5.2 Asymptotic Distribution Bounds

In this subsection we deduce the asymptotic distribution bound. The procedure is based on `Neyman-Pearson Lemma` and `Edgeworth expansion`, which is given in the following lemma.

Lemma 5.5.1 *(Zhao and Bai, 1985) Suppose that $W_1,\cdots,W_n$ are independent with mean zero and $EW_j^2>0$ and $E|W_j|^3<\infty$ for each j. Let $G_n(w)$ be the* *distribution function of the standardization of* $\sum_{i=1}^n W_i$. *Then*

$$G_n(w) = \Phi(w) + \frac{1}{6}\phi(w)(1-w^2)\mu_3\mu_2^{-3/2} + o(\frac{1}{\sqrt{n}})$$

uniformly on $w \in R^1$, *where* $\mu_2 = \sum_{i=1}^n EW_i^2$ *and* $\mu_3 = \sum_{i=1}^n EW_i^3$.

In the remainder of this subsection, we denote $J = \int \psi''(u)\psi'(u)\varphi(u)\,du$ and $K = \int\{\psi'(u)\}^3\varphi(u)\,du$. Let $P_{\beta'}$ and $E_{\beta'}$ denote probability and expectation corresponding to the parameter β', respectively.

We introduce the following assumptions.

Assumption 5.5.1 $\varphi(\cdot)$ *is three times continuously differentiable and* $\varphi^{(3)}(\cdot)$ *is a bounded function.*

Assumption 5.5.2 J *and* K *are well-defined, and* $\int \psi'''(u)\varphi(u)\,du = -3J - K$.

Assumption 5.5.3

$$\lim_{u\to\pm\infty}\psi(u) = \lim_{u\to\pm\infty}\psi'(u) = \lim_{u\to\pm\infty}\psi''(u) = 0, \ \textit{and}\ E\varepsilon^4 < \infty.$$

It will be obtained that the bound of the second order asymptotic distribution of `second order AMU` estimators of β.

Let β_0 be arbitrary but fixed point in R^1. Consider the problem of testing hypothesis

$$H^+ : \beta = \beta_1 = \beta_0 + \frac{u}{\sqrt{n}}(u > 0) \longleftrightarrow K : \beta = \beta_0.$$

Set

$$\beta_1 = \beta_0 + \Delta \text{ with } \Delta = \frac{u}{\sqrt{n}}, \text{ and } Z_{ni} = \log\frac{\varphi(Y_i - X_i\beta_0 - g(T_i))}{\varphi(Y_i - X_i\beta_1 - g(T_i))}.$$

It follows from $E\psi''(\varepsilon_i) = -I$ and $E\psi'(\varepsilon_i) = 0$ that if $\beta = \beta_1$,

$$\begin{aligned} Z_{ni} &= \log\frac{\varphi(\varepsilon_i + \Delta X_i)}{\varphi(\varepsilon_i)} \\ &= \Delta X_i\psi'(\varepsilon_i) + \frac{\Delta^2X_i^2}{2}\psi''(\varepsilon_i) + \frac{\Delta^3X_i^3}{6}\psi'''(\varepsilon_i) + \frac{\Delta^4X_i^4}{24}\psi^{(3)}(\varepsilon_i^*), \end{aligned}$$

where ε_i^* lies between ε_i and $\varepsilon_i + \Delta X_i$, and if $\beta = \beta_0$,

$$\begin{aligned} Z_{ni} &= \log\frac{\varphi(\varepsilon_i)}{\varphi(\varepsilon_i - \Delta X_i)} \\ &= \Delta X_i\psi'(\varepsilon_i) - \frac{\Delta^2X_i^2}{2}\psi''(\varepsilon_i) + \frac{\Delta^3X_i^3}{6}\psi'''(\varepsilon_i) - \frac{\Delta^4X_i^4}{24}\psi^{(3)}(\varepsilon_i^{**}), \end{aligned}$$

where ε_i^{**} lies between ε_i and $\varepsilon_i - \Delta X_i$.

We here calculate the first three moments of $\sum_{i=1}^n Z_{ni}$ at the points β_0 and β_1, respectively. Direct calculations derive the following expressions,

$$\begin{aligned} E_{\beta_0}\Big(\sum_{i=1}^n Z_{ni}\Big) &= \frac{\Delta^2 I}{2} nEX^2 - \frac{\Delta^3(3J+K)}{6} nEX^3 + o(\frac{1}{\sqrt{n}}) \\ &\stackrel{\text{def}}{=} \mu_1(\beta_0), \\ Var_{\beta_0}\Big(\sum_{i=1}^n Z_{ni}\Big) &= \Delta^2 I nEX^2 - J\Delta^3 nEX^3 + o(n\Delta^3) \\ &\stackrel{\text{def}}{=} \mu_2(\beta_0), \end{aligned}$$

$$\begin{aligned} E_{\beta_0}\sum_{i=1}^n (Z_{ni} - E_{\beta_0}Z_{ni})^3 &= E_{\beta_0}\sum_{i=1}^n Z_{ni}^3 - 3\sum_{i=1}^n E_{\beta_0}Z_{ni}^2 E_{\beta_0}Z_{ni} + 2\sum_{i=1}^n (E_{\beta_0}Z_{ni})^3 \\ &= \Delta^3 K nEX^3 + o(n\Delta^3) \\ &\stackrel{\text{def}}{=} \mu_3(\beta_0), \end{aligned}$$

$$\begin{aligned} E_{\beta_1}\Big(\sum_{i=1}^n Z_{ni}\Big) &= -\frac{\Delta^2 I}{2} nEX^2 - \frac{\Delta^3(3J+K)}{6} nEX^3 + o(\frac{1}{\sqrt{n}}) \\ &\stackrel{\text{def}}{=} \mu_1(\beta_1), \\ Var_{\beta_1}\Big(\sum_{i=1}^n Z_{ni}\Big) &= \Delta^2 I nEX^2 - J\Delta^3 nEX^3 + o(n\Delta^3) \\ &\stackrel{\text{def}}{=} \mu_2(\beta_1), \end{aligned}$$

and

$$\begin{aligned} E_{\beta_1}\sum_{i=1}^n (Z_{ni} - E_{\beta_1}Z_{ni})^3 &= E_{\beta_1}\sum_{i=1}^n Z_{ni}^3 - 3\sum_{i=1}^n E_{\beta_1}Z_{ni}^2 E_{\beta_1}Z_{ni} + 2\sum_{i=1}^n (E_{\beta_1}Z_{ni})^3 \\ &= \Delta^3 K nEX^3 + o(n\Delta^3) \\ &\stackrel{\text{def}}{=} \mu_3(\beta_1). \end{aligned}$$

First we choose an a_n such that

$$P_{\beta_1}\Big\{\sum_{i=1}^n Z_{ni} < a_n\Big\} = \frac{1}{2} + o(\frac{1}{\sqrt{n}}). \tag{5.5.1}$$

Denote

$$c_n = \frac{a_n - \sum_{i=1}^n E_{\beta_1}Z_{ni}}{\sqrt{\mu_2(\beta_1)}}, \quad d_n = \frac{a_n - \sum_{i=1}^n E_{\beta_0}Z_{ni}}{\sqrt{\mu_2(\beta_0)}}.$$

It follows from Lemma 5.5.1 that the left-hand side of (5.5.1) can be decomposed into

$$\Phi(c_n) + \phi(c_n)\frac{1-c_n^2}{6}\frac{\mu_3(\beta_1)}{\mu_2^{3/2}(\beta_1)} + o(\frac{1}{\sqrt{n}}). \tag{5.5.2}$$

This means that (5.5.1) holds if and only if $c_n = O(1/\sqrt{n})$ and $\Phi(c_n) = 1/2 + c_n\phi(c_n)$, which imply that

$$c_n = -\frac{\mu_3(\beta_1)}{6\mu_2^{3/2}(\beta_1)} + o(\frac{1}{\sqrt{n}})$$

and therefore

$$a_n = -\frac{\mu_3(\beta_1)}{6\mu_2(\beta_1)} + \sum_{i=1}^{n} E_{\beta_1}(Z_{ni}) + o(\frac{1}{\sqrt{n}}).$$

Second, we calculate $P_{\beta_0,n}\{\sum_{i=1}^n Z_{ni} \geq a_n\}$. It follows from Lemma 5.5.1 that

$$P_{\beta_0,n}\Big\{\sum_{i=1}^{n} Z_{ni} \geq a_n\Big\} = 1 - \Phi(d_n) - \phi(d_n)\frac{(1-d_n^2)\mu_3(\beta_0)}{6\mu_2^{3/2}(\beta_0)} + o(\frac{1}{\sqrt{n}}). \tag{5.5.3}$$

On the other hand, a simple calculation deduces that

$$d_n^2 = \Delta^2 InEX^2 + O(\frac{1}{\sqrt{n}}). \tag{5.5.4}$$

Substituting (5.5.4) into (5.5.3), we conclude that

$$P_{\beta_0,n}\Big\{\sum_{i=1}^{n} Z_{ni} \geq a_n\Big\} = \Phi\Big(\sqrt{\Delta^2 InEX^2}\Big) + \phi\Big(\sqrt{\Delta^2 InEX^2}\Big)$$
$$\frac{\Delta^2(3J+K)nEX^3}{6\sqrt{InEX^2}} + o(\frac{1}{\sqrt{n}}). \tag{5.5.5}$$

So far, we establish the asymptotic distribution bound, which can be summarized in the following theorem.

Theorem 5.5.1 *Assume that $EX^4 < \infty$ and that Assumptions 5.5.1-5.5.3 hold. If some estimator $\{\beta_n\}$ satisfies*

$$P_{\beta,n}\Big\{\sqrt{nIEX^2}(\beta_n - \beta) \leq u\Big\} = \Phi(u) + \phi(u)\cdot\frac{(3J+K)EX^3}{6\sqrt{nI^3E^3X^2}}u^2 + o(\frac{1}{\sqrt{n}}), \tag{5.5.6}$$

then $\{\beta_n\}$ is `second order asymptotically efficient`.

5.5.3 Construction of 2nd Order Asymptotic Efficient Estimator

In this subsection we shall construct a **second order asymptotic efficient** estimator of β. The primary estimator used here is PMLE $\widehat{\beta}_{ML}$ given in section 5.4. First of all, we study the asymptotic normality of $\widehat{\beta}_{ML}$.

Via the Taylor expansion,

$$\begin{aligned}\sum_{i=1}^n \psi(Y_i - X_i\widehat{\beta}_{ML} - \widehat{g}_P(T_i))X_i &= \sum_{i=1}^n \psi(Y_i - X_i\beta - g(T_i))X_i \\ &\quad - \sum_{i=1}^n \psi'(Y_i - X_i\beta^* - g^*(T_i))[X_i^2(\widehat{\beta}_{ML} - \beta) \\ &\quad + X_i\{\widehat{g}_P(T_i) - g(T_i)\}], \qquad (5.5.7)\end{aligned}$$

where β^* lies between $\widehat{\beta}_{ML}$ and β and $g^*(T_i)$ lies between $g(T_i)$ and $\widehat{g}_P(T_i)$.

Recalling the fact given in (5.4.1), we deduce that

$$\frac{1}{n}\sum_{i=1}^n |\psi'(Y_i - X_i\beta^* - g^*(T_i))X_i\{\widehat{g}_P(T_i) - g(T_i)\}| \to 0.$$

The definition of $\widehat{\beta}_{ML}$ implies

$$\begin{aligned}\frac{1}{\sqrt{n}}\sum_{i=1}^n \psi(Y_i - X_i\beta - g(T_i))X_i &= \frac{1}{n}\sum_{i=1}^n \psi'(Y_i - X_i\beta^* - g^*(T_i))X_i^2 \\ &\quad \sqrt{n}(\widehat{\beta}_{ML} - \beta) + o(1).\end{aligned}$$

The asymptotic normality of $\sqrt{n}(\widehat{\beta}_{ML} - \beta)$ is immediately derived and its asymptotic variance is $(IEX^2)^{-1}$.

Although we have obtained the first order asymptotic distribution of $\sqrt{n}$ $(\widehat{\beta}_{ML} - \beta)$, it is not enough for us to consider a higher order approximation. A key step is to expand the left-hand side of (5.5.7) to the second order terms. That is,

$$\begin{aligned}\sum_{i=1}^n \psi(Y_i - X_i\widehat{\beta}_{ML} - \widehat{g}_P(T_i))X_i &= \sum_{i=1}^n \psi(Y_i - X_i\beta - g(T_i))X_i \\ - \sum_{i=1}^n \psi'(Y_i - X_i\beta - g(T_i))&[X_i^2(\widehat{\beta}_{ML} - \beta) + X_i\{\widehat{g}_P(T_i) - g(T_i)\}] \\ &\quad + \frac{1}{2}\sum_{i=1}^n \psi''(Y_i - X_i\widetilde{\beta}^* - \widetilde{g}^*(T_i))X_i\{X_i(\widehat{\beta}_{ML} - \beta) \\ &\quad + \widehat{g}_P(T_i) - g(T_i)\}^2, \qquad (5.5.8)\end{aligned}$$

where $\widetilde{\beta}^*$ lies between $\widehat{\beta}_{ML}$ and β and $\widetilde{g}^*(T_i)$ lies between $g(T_i)$ and $\widehat{g}_P(T_i)$.

We introduce some notation here.

$$Z_1(\beta) = \frac{1}{\sqrt{n}}\sum_{i=1}^{n}\psi(Y_i - X_i\beta - g(T_i))X_i,$$
$$Z_2(\beta) = \frac{1}{\sqrt{n}}\sum_{i=1}^{n}\{\psi'(Y_i - X_i\beta - g(T_i))X_i^2 + EX^2 I\},$$
$$W(\beta) = \frac{1}{n}\sum_{i=1}^{n}\psi''(Y_i - X_i\widetilde{\beta}^* - \widetilde{g}^*(T_i))X_i^3.$$

Elementary calculations deduce that

$$EZ_1(\beta) = EZ_2(\beta) = 0,$$
$$EZ_1^2(\beta) = IEX^2 \stackrel{\text{def}}{=} I(\beta),$$
$$EZ_2(\beta) = E\{\psi'(\varepsilon)X^2 + EX^2 I\}^2,$$
$$EZ_1(\beta)Z_2(\beta) = E\psi(\varepsilon)\psi'(\varepsilon)EX^3 = JEX^3.$$

A central limit theorem implies that $Z_1(\beta)$ and $Z_2(\beta)$ have asymptotically normal distributions with mean zero, variances $I(\beta)$ and $E\{\psi'(\varepsilon)X^2 + EX^2 I\}^2 (\stackrel{\text{def}}{=} L(\beta))$, respectively and covariance $J(\beta) = JEX^3$, while the law of large numbers implies that $W(\beta)$ converges to $E\psi''(\varepsilon)EX^3 = -(3J + K)EX^3$. Combining these arguments with (5.5.8), we obtain an asymptotic expansion of $\sqrt{n}(\widehat{\beta}_{ML} - \beta)$. That is,

$$\sqrt{n}(\widehat{\beta}_{ML} - \beta) = \frac{Z_1(\beta)}{I(\beta)} + \frac{Z_1(\beta)Z_2(\beta)}{\sqrt{n}I^2(\beta)} - \frac{(3J+K)EX^3}{2\sqrt{n}I^3(\beta)}Z_1^2(\beta) + o_p\Big(\frac{1}{\sqrt{n}}\Big). \tag{5.5.9}$$

Further study indicates that $\widehat{\beta}_{ML}$ is not second order asymptotic efficient. However, we have the following result.

Theorem 5.5.2 *Suppose that the m-th($m \geq 4$) cumulants of $\sqrt{n}\widehat{\beta}_{ML}$ are less than order $1/\sqrt{n}$ and that Assumptions 5.5.1- 5.5.3 hold. Then*

$$\widehat{\beta}_{ML}^* = \widehat{\beta}_{ML} + \frac{KEX^3}{3nI^2(\beta)}$$

is second order asymptotically efficient.

Proof. Denote $\widehat{T}_n = \sqrt{n}(\widehat{\beta}_{ML} - \beta)$. We obtain from (5.5.9) and Assumptions 5.5.1-5.5.3 that

$$E_\beta \widehat{T}_n = -\frac{(J+K)EX^3}{2\sqrt{n}I^2(\beta)} + o\Big(\frac{1}{\sqrt{n}}\Big),$$

$$\begin{aligned} Var_\beta \widehat{T}_n &= \frac{1}{I(\beta)} + o(\frac{1}{\sqrt{n}}), \\ E_\beta(\widehat{T}_n - E_\beta \widehat{T}_n)^3 &= -\frac{(3J+K)EX^3}{\sqrt{n}I^3(\beta)} + o(\frac{1}{\sqrt{n}}). \end{aligned}$$

Denote $u(\beta) = -\frac{KEX^3}{3I^2(\beta)}$. The cumulants of the asymptotic distribution of $\sqrt{nI(\beta)}(\widehat{\beta}^*_{ML} - \beta)$ can be approximated as follows:

$$\begin{aligned} &\sqrt{nI(\beta)}E_\beta(\widehat{\beta}^*_{ML} - \beta) = -\sqrt{\frac{I(\beta)}{n}}\Big\{\frac{(J+K)EX^3}{2I^2(\beta)} + u(\beta)\Big\} + o(\frac{1}{\sqrt{n}}), \\ &nI(\beta)Var_\beta(\widehat{\beta}^*_{ML}) = 1 + o(\frac{1}{\sqrt{n}}), \\ &E_\beta\{\sqrt{nI(\beta)}(\widehat{\beta}^*_{ML} - E_\beta\widehat{\beta}^*_{ML})\}^3 = -\frac{(3J+K)EX^3}{\sqrt{n}I^{3/2}(\beta)} + o(\frac{1}{\sqrt{n}}). \end{aligned}$$

Obviously $\widehat{\beta}^*_{ML}$ is an AMU estimator. Moreover, using Lemma 5.5.1 again and the same technique as in the proof of Theorem 5.5.1, we deduce that

$$P_{\beta,n}\{\sqrt{nI(\beta)}(\widehat{\beta}^*_{ML} - \beta) \le t\} = \Phi(t) + \phi(t)\frac{(3J+K)EX^3}{6\sqrt{nI^3(\beta)}}t^2 + o(\frac{1}{\sqrt{n}}). \quad (5.5.10)$$

The proof of Theorem 5.5.2 is finished by combining (5.5.6) and (5.5.10).

5.6 Estimation of the Error Distribution

5.6.1 Introduction

This section discusses asymptotic behaviors of the estimator of the error density function $\varphi(u)$, $\widehat{\varphi}_n(u)$, which is defined by using the estimators given in (1.2.2) and (1.2.3). Under appropriate conditions, we first show that $\widehat{\varphi}_n(u)$ converges in probability, almost surely converges and uniformly almost surely converges. Then we consider the asymptotic normality and the convergence rates of $\widehat{\varphi}_n(u)$. Finally we establish the LIL for $\widehat{\varphi}_n(u)$.

Set $\widehat{\varepsilon}_i = Y_i - X_i^T\beta_{LS} - \widehat{g}_n(T_i)$ for $i = 1, \ldots, n$. Define the estimator of $\varphi(u)$ as follows,

$$\widehat{\varphi}_n(u) = \frac{1}{2na_n}\sum_{i=1}^n I_{(u-a_n \le \widehat{\varepsilon}_i \le u+a_n)}, \quad u \in R^1 \quad (5.6.1)$$

where $a_n (> 0)$ is a bandwidth and I_A denotes the indicator function of the set A. Estimator $\widehat{\varphi}_n$ is a special form of the general nonparametric density estimation.

5.6.2 Consistency Results

In this subsection we shall consider the case where ε_i are i.i.d. and (X_i, T_i) are fixed design points, and prove that $\widehat{\varphi}_n(u)$ converges in probability, almost surely converges and uniformly almost surely converges. In the following, we always denote

$$\varphi_n(u) = \frac{1}{2na_n}\sum_{i=1}^{n} I_{(u-a_{\le}\varepsilon_i \le u+a_n)}$$

for fixed point $u \in C(\varphi)$, where $C(\varphi)$ is the set of the continuous points of φ.

Theorem 5.6.1 *There exists a constant $M > 0$ such that $\|X_i\| \le M$ for $i = 1, \cdots, n$. Assume that Assumptions 1.3.1-1.3.3 hold. Let*

$$0 < a_n \to 0 \text{ and } n^{1/3} a_n \log^{-1} n \to \infty.$$

Then $\widehat{\varphi}_n(u) \to \varphi(u)$ in probability as $n \to \infty$.

Proof. A simple calculation shows that the mean of $\varphi_n(u)$ converges to $\varphi(u)$ and its variance converges to 0. This implies that $\varphi_n(u) \to \varphi(u)$ in probability as $n \to \infty$.

Now, we prove $\widehat{\varphi}_n(u) - \varphi_n(u) \to 0$ in probability.

If $\varepsilon_i < u - a_n$, then $\widehat{\varepsilon}_i \in (u - a_n, u + a_n)$ implies $u - a_n + X_i^T(\beta_{LS} - \beta) + \widehat{g}_n(T_i) - g(T_i) < \varepsilon_i < u - a_n$. If $\varepsilon_i > u + a_n$, then $\widehat{\varepsilon}_i \in (u - a_n, u + a_n)$ implies $u + a_n < \varepsilon_i < u + a_n + X_i^T(\beta_{LS} - \beta) + \widehat{g}_n(T_i) - g(T_i)$. Write

$$C_{ni} = X_i^T(\beta_{LS} - \beta) + \widehat{g}_n(T_i) - g(T_i) \text{ for } i = 1, \ldots, n.$$

It follows from (2.1.2) that, for any $\zeta > 0$, there exists a $\eta_0 > 0$ such that

$$P\{n^{1/3} \log^{-1} n \sup_i |C_{ni}| > \eta_0\} \le \zeta.$$

The above arguments yield that

$$\begin{aligned} |\widehat{\varphi}_n(u) - \varphi_n(u)| &\le \frac{1}{2na_n} I_{(u \pm a_n - |C_{ni}| \le \varepsilon_i \le u \pm a_n)} + \frac{1}{2na_n} I_{(u \pm a_n \le \varepsilon_i \le u \pm a_n + |C_{ni}|)} \\ &\stackrel{\text{def}}{=} I_{1n} + I_{2n}, \end{aligned}$$

where $I_{(u \pm a_n - |C_{ni}| \le \varepsilon_i \le u \pm a_n)}$ denotes $I_{(u + a_n - |C_{ni}| \le \varepsilon_i \le u + a_n) \cup (u - a_n - |C_{ni}| \le \varepsilon_i \le u - a_n)}$.

We shall complete the proof of the theorem by dealing with I_{1n} and I_{2n}. For any $\zeta' > 0$ and large enough n,

$$\begin{aligned} P\{I_{1n} > \zeta'\} &\le \zeta + P\{I_{1n} > \zeta', \sup_i |C_{ni}| \le \eta_0\} \\ &\le \zeta + P\Big\{\sum_{i=1}^n I_{(u\pm a_n - C\eta_0 n^{-1/3}\log n \le \varepsilon_i \le u\pm a_n)} \ge 2na_n\zeta'\Big\}. \end{aligned}$$

Using the continuity of φ on u and applying Chebyshev's inequality we know that the second term is smaller than

$$\begin{aligned} &\frac{1}{2a_n\zeta'}P\Big(u \pm a_n - C\eta_0 n^{-1/3}\log n \le \varepsilon_i \le u \pm a_n\Big) \\ &\qquad\qquad = \frac{C\eta_0 \log n}{2\zeta' n^{1/3} a_n}\{\varphi(u) + o(1)\}. \end{aligned}$$

It follows from $a_n n^{1/3}\log^{-1} n \to \infty$ that

$$\limsup_{n\to\infty} P\{I_{1n} > \zeta'\} \le \zeta.$$

Since ζ is arbitrary, we obtain $I_{1n} \to 0$ in probability as $n \to \infty$. We can similarly prove that I_{2n} tends to zero in probability as $n \to \infty$. Thus, we complete the proof of Theorem 5.6.1.

Theorem 5.6.2 *Assume that the conditions of Theorem 5.6.1 hold. Furthermore,*

$$n^{1/3} a_n \log^{-2} n \to \infty. \tag{5.6.2}$$

Then $\widehat{\varphi}_n(u) \to \varphi(u)$ for $u \in C(\varphi)$ a.s. as $n \to \infty$.

Proof. Set $\varphi_n^E(u) = E\varphi_n(u)$ for $u \in C(\varphi)$. Using the continuity of φ on u and $a_n \to 0$, it can be shown that

$$\varphi_n^E(u) \to \varphi(u) \quad \text{as } n \to \infty. \tag{5.6.3}$$

Consider $\varphi_n(u) - \varphi_n^E(u)$, which can be represented as

$$\begin{aligned} \varphi_n(u) - \varphi_n^E(u) &= \frac{1}{2na_n}\sum_{i=1}^n\Big\{I_{(u-a_n\le\varepsilon_i\le u+a_n)} - EI_{(u-a_n\le\varepsilon_i\le u+a_n)}\Big\} \\ &\overset{\text{def}}{=} \frac{1}{2na_n}\sum_{i=1}^n U_{ni}. \end{aligned}$$

$U_{n1},\ldots,U_{nn}$ are then independent with mean zero and $|U_{ni}|\le 1$, and $\mathrm{Var}(U_{ni})\le P(u-a_n\le\varepsilon_i\le u+a_n)=2a_n\varphi(u)(1+o(1))\le 4a_n\varphi(u)$ for large enough n. It follows from **Bernstein's inequality** that for any $\zeta>0$,

$$\begin{aligned} P\{|\varphi_n(u)-\varphi_n^E(u)|\ge\zeta\} &= P\Big(\Big|\sum_{i=1}^n U_{ni}\Big|\ge 2na_n\zeta\Big)\\ &\le 2\exp\Big\{-\frac{4n^2a_n^2\zeta^2}{8na_n\varphi(u)+4/3na_n\zeta}\Big\}\\ &= 2\exp\Big\{-\frac{3na_n\zeta^2}{6\varphi(u)+\zeta}\Big\}. \end{aligned} \tag{5.6.4}$$

Condition (5.6.2) and Borel-Cantelli Lemma imply

$$\varphi_n(u)-\varphi_n^E(u)\to 0\quad a.s. \tag{5.6.5}$$

In the following, we shall prove

$$\widehat{\varphi}_n(u)-\varphi_n(u)\to 0\quad a.s. \tag{5.6.6}$$

According to (2.1.2), we have with probability one that

$$\begin{aligned} |\widehat{\varphi}_n(u)-\varphi_n(u)| &\le \frac{1}{2na_n}I_{(u\pm a_n-Cn^{-1/3}\log n\le\varepsilon_i\le u\pm a_n)}\\ &\quad+\frac{1}{2na_n}I_{(u\pm a_n\le\varepsilon_i\le u\pm a_n+Cn^{-1/3}\log n)}\\ &\stackrel{\text{def}}{=} J_{1n}+J_{2n}. \end{aligned} \tag{5.6.7}$$

Denote

$$\varphi_{n1}(u)=\frac{1}{2a_n}P(u\pm a_n-Cn^{-1/3}\log n\le\varepsilon_i\le u\pm a_n). \tag{5.6.8}$$

Then $\varphi_{n1}(u)\le C\varphi(u)(n^{1/3}a_n)^{-1}\log n$ for large enough n. By Condition (5.6.2), we obtain

$$\varphi_{n1}(u)\to 0\quad\text{as } n\to\infty. \tag{5.6.9}$$

Now let us deal with $J_{n1}-\varphi_{n1}(u)$. Set

$$Q_{ni}=I_{(u\pm a_n-Cn^{-1/3}\log n\le\varepsilon_i\le u\pm a_n)}-P(u\pm a_n-Cn^{-1/3}\log n\le\varepsilon_i\le u\pm a_n),$$

for $i=1,\ldots,n$. Then $Q_{n1},\ldots,Q_{nn}$ are independent with mean zero and $|Q_{ni}|\le 1$, and $\mathrm{Var}(Q_{ni})\le 2Cn^{-1/3}\log n\varphi(u)$. By **Bernstein's inequality**, we have

$$\begin{aligned} P\{|J_{n1}-\varphi_{n1}(u)|>\zeta\} &= P\Big\{\Big|\sum_{i=1}^n Q_{ni}\Big|>\zeta\Big\}\\ &\le 2\exp\Big\{-\frac{Cna_n\zeta^2}{n^{-1/3}a_n^{-1}\varphi(u)\log^{-1}n+\zeta}\Big\}\\ &\le 2\exp(-Cna_n\zeta). \end{aligned} \tag{5.6.10}$$

Equation (5.6.10) and Borel-Cantelli Lemma imply that

$$J_{n1} - \varphi_{n1}(u) \to 0, \quad a.s.$$

Combining (5.6.9) with the above conclusion, we obtain $J_{n1} \to 0$ a.s. Similar arguments yield $J_{n2} \to 0$ a.s. From (5.6.3), (5.6.5) and (5.6.6), we complete the proof of Theorem 5.6.2.

Theorem 5.6.3 *Assume that the conditions of Theorem 5.6.2 hold. In addition, φ is uniformly continuous on R^1. Then* $\sup_u |\widehat{\varphi}_n(u) - \varphi(u)| \to 0$, *a.s.*

We need the following conclusion to prove Theorem 5.6.3.

Conclusion D. (See Devroye and Wagner, 1980) *Let μ_n and μ be 1-dimensional empirical distribution and theoretical distribution, respectively, $a > 0$ and Ia be an interval with length a. Then for any $\zeta > 0$, $0 < b \le 1/4$ and $n \ge \max\{1/b, 8b/\zeta^2\}$,*

$$P\Big(\sup\{|\mu_n(Ia) - \mu(Ia)| : 0 < \mu(Ia) \le b\} \ge \zeta\Big) \le 16n^2 \exp\{-n\zeta^2/(64b + 4\zeta)\} \\ +8n\exp\{-nb/10\}.$$

Proof of Theorem 5.6.3. We still use the notation in the proof of Theorem 5.6.2 to denote the empirical distribution of $\varepsilon_1, \ldots, \varepsilon_n$ by μ_n and the distribution of ε by μ. Since φ is uniformly continuous, we have $\sup_u \varphi(u) = \varphi_0 < \infty$. It is easy to show that

$$\sup_u |\varphi(u) - f_n^E(u)| \to 0 \quad \text{as } n \to \infty. \tag{5.6.11}$$

Write

$$\varphi_n(u) - f_n^E(u) = \frac{1}{2a_n}\{\mu_n([u - a_n, u + a_n]) - \mu([u - a_n, u + a_n])\}.$$

and denote $b_n^* = 2\varphi_0 a_n$ and $\zeta_n = 2a_n\zeta$ for any $\zeta > 0$. Then for large enough n, $0 < b_n^* < 1/4$ and $\sup_u \mu([u - a_n, u + a_n]) \le b_n^*$ for all n. From Conclusion D, we have for large enough n

$$\begin{aligned} P\{\sup_u |\varphi_n(u) - \varphi_n^E(u)| \ge \zeta\} &= P\{\sup_u |\mu_n([u - a_n, u + a_n]) \\ &\quad -\mu([u - a_n, u + a_n])| \ge 2a_n\zeta\} \\ &\le 16n^2 \exp\Big\{-\frac{na_n^2\zeta^2}{32\varphi_0 a_n + 2a_n\zeta}\Big\} \\ &\quad +8n\exp\Big\{-\frac{na_n^2\varphi_0}{5}\Big\}. \end{aligned}$$

It follows from (5.6.2) and Borel-Cantelli Lemma that

$$\sup_u |\varphi_n(u) - \varphi_n^E(u)| \to 0 \ \ a.s. \tag{5.6.12}$$

Combining (5.6.12) with (5.6.11), we obtain

$$\sup_u |\varphi_n(u) - \varphi(u)| \to 0 \quad \text{a.s.} \tag{5.6.13}$$

In the following we shall prove that

$$\sup_u |\widehat{\varphi}_n(u) - \varphi_n(u)| \to 0 \quad \text{a.s.} \tag{5.6.14}$$

It is obvious that $\sup_u |\varphi_{n1}(u)| \to 0$ as $n \to \infty$. Set $d_n = \varphi_0 n^{-1/3} \log n$. For large enough n, we have $0 < d_n < 1/4$ and

$$\sup_u \mu\{(u \pm a_n - Cn^{-1/3} \log n, u \pm a_n)\} \leq C d_n \text{ for all } n.$$

It follows from Conclusion D that

$$\begin{aligned} P(\sup_u |J_{n1} - \varphi_{n1}(u)| > \zeta) &\leq P\Big(|\mu_n\{(u \pm a_n - Cn^{-1/3}\log n, u \pm a_n)\} \\ &\qquad -\mu\{(u \pm a_n - Cn^{-1/3}\log n, u \pm a_n)\}| \geq 2a_n\zeta\Big) \\ &\leq 16n^2 \exp\Big(-\frac{na_n^2\zeta^2}{16\varphi_0 n^{-1/3}\log n + 2a_n\zeta}\Big) \\ &\qquad +8n\exp\Big(-\frac{n^{2/3}\log n}{10}\Big), \end{aligned}$$

which and (5.6.2) imply that $\sup_u |J_{n1} - \varphi_{n1}(u)| \to 0$ a.s. and hence $\sup_u |J_{n1}| \to 0$ a.s. We can prove $\sup_u |J_{n2}| \to 0$ similarly. Recalling the proof of Theorem 5.6.2, we can show that for large enough n

$$\sup_u |\widehat{\varphi}_n(u) - \varphi_n(u)| \leq \sup_u |J_{n1}| + \sup_u |J_{n2}|$$

with probability one, which implies (5.6.14) and the conclusion of Theorem 5.6.3 follows.

5.6.3 Convergence Rates

Theorem 5.6.4 *Assume that the conditions of Theorem 5.6.2 hold. If φ is locally* `Lipschitz continuous` *of order 1 on u. Then taking $a_n = n^{-1/6}\log^{1/2} n$, we have*

$$\widehat{\varphi}_n(u) - \varphi(u) = O(n^{-1/6}\log^{1/2} n), \quad a.s. \tag{5.6.15}$$

Proof. The proof is analogous to that of Theorem 5.6.2. By the conditions of Theorem 5.6.4, there exist $c_0 > 0$ and $\delta_1 = \delta_1(u) > 0$ such that $|\varphi(u') - \varphi(u)| \le c_0|u' - u|$ for $u' \in (u - \delta_1, u + \delta_1)$. Hence for large enough n

$$\begin{aligned} |\varphi_n^E(u) - \varphi(u)| &\le \frac{1}{2a_n}\int_{u-a_n}^{u+a_n} |\varphi(u) - \varphi(u')|du' \\ &\le c_0 a_n/2 = O(n^{-1/6}\log^{1/2} n). \end{aligned} \tag{5.6.16}$$

Since φ is bounded on $(u - \delta_1, u + \delta_1)$, we have for large enough n

$$\begin{aligned} \varphi_{n1}(u) &= \frac{1}{2a_n}P(u \pm a_n - Cn^{-1/3}\log n \le \varepsilon_i \le u \pm a_n) \\ &\le Cn^{-1/3}a_n^{-1}\log n \sup_{u'\in(u-\delta_1,u+\delta_1)} \varphi(u') \\ &= O(n^{-1/6}\log^{1/2} n). \end{aligned}$$

Replacing ζ by $\zeta_n = \zeta n^{-1/6}\log^{1/2} n$ in (5.6.4), then for large enough n

$$P\{|\varphi_n(u) - \varphi_n^E(u)| \ge 2\zeta n^{-1/6}\log^{1/2} n\} \le 2\exp\left\{-\frac{3n^{1/2}\log^{3/2} n\zeta}{6\varphi_0 + \zeta}\right\},$$

here $\varphi_0 = \sup_{u'\in(u-\delta_1,u+\delta_1)}\varphi(u')$. Instead of (5.6.12), we have

$$\varphi_n(u) - \varphi_n^E(u) = O(n^{-1/6}\log^{1/2} n), \quad a.s. \tag{5.6.17}$$

The same argument as (5.6.10) yields

$$P\{|J_{n1} - \varphi_{n1}(u)| > \zeta n^{-1/6}\log^{1/2} n\} \le 2\exp(-Cn^{2/3}\log^{1/2} n).$$

Hence, $J_{n1} - \varphi_{n1}(u) = O(n^{-1/6}\log^{1/2} n)$ a.s. Equations (5.6.16) and (5.6.17) imply

$$\varphi_n(u) - \varphi(u) = O(n^{-1/6}\log^{1/2} n), \quad a.s.$$

This completes the proof of Theorem 5.6.4.

5.6.4 Asymptotic Normality and LIL

Theorem 5.6.5 *Assume that the conditions of Theorem 5.6.2 hold. In addition, φ is locally* Lipschitz continuous *of order 1 on u. Let* $\lim_{n\to\infty} na_n^3 = 0$. *Then* $\sqrt{2na_n/\varphi(u)}\{\widehat{\varphi}_n(u) - \varphi(u)\} \longrightarrow^{\mathcal{L}} N(0,1)$.

Theorem 5.6.6 *Assume that the conditions of Theorem 5.6.2 hold. In addition, φ is locally* Lipschitz continuous *of order 1 on u, Let* $\lim_{n\to\infty} na_n^3/\log\log n = 0$. *Then*

$$\limsup_{n\to\infty} \pm\Big\{\frac{na_n}{\varphi(u)\log\log n}\Big\}^{1/2}\{\widehat{\varphi}_n(u) - \varphi(u)\} = 1, \quad a.s.$$

The proofs of the above two theorems can be completed by slightly modifying the proofs of Theorems 2 and 3 of Chai and Li (1993). Here we omit the details.

6
PARTIALLY LINEAR TIME SERIES MODELS

6.1 Introduction

Previous chapters considered the partially linear models in the framework of independent observations. The independence assumption can be reasonable when data are collected from a human population or certain measurements. However, there are classical data sets such as the sunspot, lynx and the Australian blowfly data where the independence assumption is far from being fulfilled. In addition, recent developments in semiparametric regression have provided a solid foundation for partially time series analysis. In this chapter, we pay attention to `partially linear time series` models and establish asymptotic results as well as small sample studies.

The organization of this chapter is as follows. Section 6.2 presents several data-based `test statistics` to determine which model should be chosen to model a partially linear dynamical system. Section 6.3 proposes a cross-validation (CV) based criterion to select the optimum linear subset for a partially linear regression model. In Section 6.4, we investigate the problem of selecting optimum smoothing parameter for a `partially linear autoregressive model`. Section 6.5 summarizes recent developments in a general class of `additive stochastic regression models`.

6.2 Adaptive Parametric and Nonparametric Tests

6.2.1 Asymptotic Distributions of Test Statistics

Consider a partially linear dynamical model of the form

$$Y_t = U_t^T \beta + g(V_t) + e_t, 1 \le t \le T \tag{6.2.1}$$

where T is the number of observations, $\beta = (\beta_1, \ldots, \beta_p)^T$ is a vector of unknown parameters, g is an unknown and nonlinear function over R^d, $U_t = (U_{t1}, \ldots, U_{tp})^T$ and $V_t = (V_{t1}, \ldots, V_{td})^T$ are random processes, $X_t = (U_t^T, V_t^T)^T$, (X_t, Y_t) are strictly `stationary processes`, $\{e_t\}$ is i.i.d. error processes with $Ee_t = 0$ and $0 < Ee_t^2 = \sigma_0^2 < \infty$, and the $\{e_s\}$ is independent of $\{X_t\}$ for all $s \ge t$.

For identifiability, we assume that the (β, g) satisfies

$$E\{Y_t - U_t^T\beta - g(V_t)\}^2 = \min_{(\alpha,f)} E\{Y_t - U_t^T\alpha - f(V_t)\}^2$$

For the case where $\{X_t\}$ is a sequence of i.i.d. random variables, model (6.2.1) with $d = 1$ has been discussed in the previous chapters. For $Y_t = y_{t+p+d}$, $U_{ti} = y_{t+p+d-i}$ $(1 \le i \le p)$ and $V_{tj} = y_{t+p+d-j}$ $(1 \le j \le d)$, model (6.2.1) is a semiparametric AR model of the form

$$y_{t+p+d} = \sum_{i=1}^{p} y_{t+p+d-i}\beta_i + g(y_{t+p+d-1}, \ldots, y_{t+p}) + e_t. \tag{6.2.2}$$

This model was first discussed by Robinson (1988). Recently, Gao and Liang (1995) established the asymptotic normality of the least squares estimator of β. See also Gao (1998) and Liang (1996) for some other results. For $Y_t = y_{t+p+d}$, $U_{ti} = y_{t+p+d-i}$ $(1 \le i \le p)$ and $\{V_t\}$ is a vector of exogenous variables, model (6.2.1) is an additive ARX model of the form

$$y_{t+p+d} = \sum_{i=1}^{p} y_{t+p+d-i}\beta_i + g(V_t) + e_t. \tag{6.2.3}$$

See Chapter 48 of Teräsvirta, Tjøstheim and Granger (1994) for more details. For the case where both U_t and V_t are stationary AR time series, model (6.2.1) is an additive state-space model of the form

$$\left.\begin{aligned} Y_t &= U_t^T\beta + g(V_t) + e_t, \\ U_t &= f(U_{t-1}) + \delta_t, \\ V_t &= h(V_{t-1}) + \eta_t, \end{aligned}\right\} \tag{6.2.4}$$

where the functions f and h are smooth functions, and the δ_t and η_t are error processes.

In this section, we consider the case where p is a finite integer or $p = p_T \to \infty$ as $T \to \infty$. By approximating $g(\cdot)$ by an `orthogonal series` of the form $\sum_{i=1}^{q} z_i(\cdot)\gamma_i$, where $q = q_T$ is the number of summands, $\{z_i(\cdot) : i = 1, 2, \ldots\}$ is a prespecified family of orthogonal functions and $\gamma = (\gamma_1, \ldots, \gamma_q)^T$ is a vector of unknown parameters, we define the least squares estimator $(\widehat{\beta}, \widehat{\gamma})$ of (β, γ) as the solution of

$$\sum_{t=1}^{T} \{Y_t - U_t^T\widehat{\beta} - Z(V_t)^T\widehat{\gamma}\}^2 = \min!, \tag{6.2.5}$$

where $Z(\cdot) = Z_q(\cdot) = \{z_1(\cdot), \ldots, z_q(\cdot)\}^T$.

It follows from (6.2.5) that

$$\widehat{\beta} = (\widehat{U}^T\widehat{U})^+\widehat{U}^TY, \tag{6.2.6}$$

$$\widehat{\gamma} = (Z^TZ)^+Z^T\{\mathcal{F} - U(\widehat{U}^T\widehat{U})^+\widehat{U}^T\}Y, \tag{6.2.7}$$

where $Y = (Y_1, \ldots, Y_T)^T$, $U = (U_1, \ldots, U_T)^T$, $Z = \{Z(V_1), \ldots, Z(V_T)\}^T$, $P = Z(Z^TZ)^+Z^T$, $\widehat{U} = (\mathcal{F} - P)U$, and $(\cdot)^+$ denotes the Moore-Penrose inverse.

Thus, the corresponding nonparametric estimator can be defined by

$$\widehat{g}(v) = Z(v)^T\widehat{\gamma}. \tag{6.2.8}$$

Given the data $\{(U_t, V_t, Y_t) : 1 \le t \le T\}$, our objective is to test whether a partially linear model of the form

$$Y_t = U_t^T\beta + g(V_t) + e_t$$

is better than either $Y_t = U_t^T\beta + e_t$ or $Y_t = g(V_t) + e_t$. which is equivalent to testing the hypothesis $H_{0g} : g = 0$ or $H_{0\beta} : \beta = 0$. This suggests using a statistic of the form

$$L_{1T} = (2q)^{-1/2}\sigma_0^{-2}\{\widehat{\gamma}^TZ^TZ\widehat{\gamma} - q\sigma_0^2\} \tag{6.2.9}$$

for testing H_{0g} or a statistic of the form

$$L_{2T} = (2p)^{-1/2}\sigma_0^{-2}\{\widehat{\beta}^TU^TU\widehat{\beta} - p\sigma_0^2\},$$

for testing $H_{0\beta}$.

Now, we have the main results of this section.

Theorem 6.2.1 *(i) Assume that Assumptions 6.6.1-6.6.4 listed in Section 6.6 hold. If* $g(V_t) = q^{1/4}/\sqrt{T}g_0(V_t)$ *with* $g_0(V_t)$ *satisfying* $0 < E\{g_0(V_t)^2\} < \infty$, *then as* $T \to \infty$

$$L_{1T} \longrightarrow^{\mathcal{L}} N(L_{10}, 1) \tag{6.2.10}$$

where $L_{10} = (\sqrt{2}\sigma_0^2)^{-1}E\{g_0(V_t)^2\}$. *Furthermore, under* $H_{1g} : g \ne 0$, *we have* $\lim_{T\to\infty} P(L_{1T} \ge C_{1T}) = 1$, *where* C_{1T} *is any positive, nonstochastic sequence with* $C_{1T} = o(Tq^{-1/2})$.

(ii) Assume that Assumptions 6.6.1-6.6.4 listed in Section 6.6 hold. If $\beta = p^{1/4}/\sqrt{T}\beta_0$ *with* $0 < E(U_t^T\beta_0)^2 < \infty$, *then as* $T \to \infty$

$$L_{2T} \longrightarrow^{\mathcal{L}} N(L_{20}, 1) \tag{6.2.11}$$

where $L_{20} = (\sqrt{2}\sigma_0^2)^{-1}E(U_t^T\beta_0)^2$. *Furthermore, under* $H_{1\beta} : \beta \neq 0$, *we have* $\lim_{T\to\infty} P(L_{2T} \geq C_{2T}) = 1$, *where* C_{2T} *is any positive, nonstochastic sequence with* $C_{2T} = o(Tp^{-1/2})$.

Let $L_T = L_{1T}$ or L_{2T} and H_0 denote H_{0g} or $H_{0\beta}$. It follows from (6.2.10) or (6.2.11) that L_T has an asymptotic normality distribution under the null hypothesis H_0. In general, H_0 should be rejected if L_T exceeds a critical value, L_0^*, of normal distribution. The proof of Theorem 6.2.1 is given in Section 6.6. Power investigations of the test statistics are reported in Subsection 6.2.2.

Remark 6.2.1 *Theorem 6.2.1 provides the test statistics for testing the partially linear dynamical model (6.2.1). The test procedures can be applied to determine a number of models including (6.2.2)–(6.2.4) (see Examples 6.2.1 and 6.2.2 below). Similar discussions for the case where the observations in (6.2.1) are i.i.d. have already been given by several authors (see Eubank and Spiegelman (1990), Fan and Li (1996), Jayasuriva (1996), Gao and Shi (1995) and Gao and Liang (1997)). Theorem 6.2.1 complements and generalizes the existing discussions for the i.i.d. case.*

Remark 6.2.2 *In this section, we consider model (6.2.1). For the sake of identifiability, we need only to consider the following transformed model*

$$Y_t = \beta_0 + \sum_{i=1}^{p} \widetilde{U}_{ti}\beta_i + \widetilde{g}(V_t) + e_t,$$

where $\beta_0 = \sum_{i=1}^{p} E(U_{ti})\beta_i + Eg(V_t)$ *is an unknown parameter,* $\widetilde{U}_{ti} = U_{ti} - EU_{ti}$ *and* $\widetilde{g}(V_t) = g(V_t) - Eg(V_t)$. *It is obvious from the proof in Section 6.6 that the conclusion of Theorem 6.2.1 remains unchanged when* Y_t *is replaced by* $\widetilde{Y}_t = Y_t - \widehat{\beta}_0$, *where* $\widehat{\beta}_0 = 1/T\sum_{t=1}^{T} Y_t$ *is defined as the estimator of* β_0.

Remark 6.2.3 *In this chapter, we choose the traditional LS estimation method. However, it is well known that the estimators based on the LS method are sensitive to outliers and that the error distribution may be heavy-tailed. Thus, a* *more robust estimation procedure for the nonparametric component $g(\cdot)$ might be worthwhile to study in order to achieve desirable robustness properties. A recent paper by Gao and Shi (1995) on M-type smoothing splines for nonparametric and semiparametric regression can be used to construct a* `test statistic` *based on the following M-type estimator $\widehat{g}(\cdot) = Z(\cdot)^T \widehat{\gamma}_M$,*

$$\sum_{t=1}^{T} \rho\{Y_t - U_t^T \widehat{\beta}_M - Z(V_t)^T \widehat{\gamma}_M\} = \min!,$$

where $\rho(\cdot)$ is a convex function.

Remark 6.2.4 *The construction of the* `test statistic` *(6.2.9) is based on the fact that g is approximated by the* `orthogonal series`*. The inverse matrix $(Z^T Z)^{-1}$ involved in the* `test statistic` *(6.2.9) is just a random matrix of $q \times q$ order. We can estimate g by a kernel estimator and construct a kernel-based* `test statistic` *for testing $H_{0g} : g = 0$. The proof of the asymptotic normality of the kernel-based statistic is much more complicated than that of Theorem 6.2.1(i) due to the fact that a random inverse matrix of $T \times T$ order is involved in the kernel-based statistic. More recently, Kreiss, Neumann and Yao (1997) avoided using this kind of* `test statistic` *by adopting an alternative version.*

Remark 6.2.5 *Consider the case where $\{e_t\}$ is a sequence of long-range dependent error processes given by*

$$e_t = \sum_{s=0}^{\infty} b_s \varepsilon_{t-s}$$

with $\sum_{s=0}^{\infty} b_s^2 < \infty$, where $\{\varepsilon_s\}$ is a sequence of i.i.d. random processes with mean zero and variance one. More recently, Gao and Anh (1999) have established a result similar to Theorem 6.2.1.

6.2.2 Power Investigations of the Test Statistics

In this section, we illustrate Theorem 6.2.1 by a number of simulated and real examples. Rejection rates of the `test statistics` to test linearity and additivity are detailed in Examples 6.2.1 and 6.2.2 respectively. Both linearity and additivity are also demonstrated by a real example.

Example 6.2.1 *Consider an ARX model of the form*

$$\left.\begin{aligned} y_t &= 0.25 y_{t-1} + \delta x_t^2 + e_t, \\ x_t &= 0.5 x_{t-1} + \varepsilon_t, \end{aligned}\right\} \quad 1 \le t \le T, \tag{6.2.12}$$

where $0 \leq \delta \leq 1$ *is a constant,* e_t *and* ε_t *are mutually independent and identically distributed random errors,* $e_t \sim N(0, \sigma_0^2)$, $\varepsilon_t \sim U(-0.5, 0.5)$, x_0 *is independent of* y_0, $x_0 \sim U(-1/3, 1/3)$, $y_0 \sim N(\mu_1, \sigma_1^2)$, e_s *and* ε_t *are independent for all* s *and* t, (ε_t, e_t) *are independent of* (x_0, y_0), *and the parameters* σ_0, μ_1 *and* σ_1 *are chosen such that* (x_t, y_t) *are stationary.*

Example 6.2.2 *Consider a state-space model of the form*

$$\begin{aligned} y_t &= \phi u_t + 0.75 v_t^2 + e_t, \\ u_t &= 0.5 u_{t-1} + \varepsilon_t, \qquad 1 \leq t \leq T, \\ v_t &= 0.5 \frac{v_{t-1}}{1 + v_{t-1}^2} + \eta_t, \end{aligned}$$

where $0 \leq \phi \leq 1$ *is a constant, both* $\{\varepsilon_t : t \geq 1\}$ *and* $\{\eta_t : t \geq 1\}$ *are mutually independent and identically distributed,* $\{\varepsilon_t : t \geq 1\}$ *is independent of* u_0, $\{v_t : t \geq 1\}$ *is independent of* v_0, $\varepsilon_t \sim U(-0.5, 0.5)$, $u_0 \sim U(-1/3, 1/3)$, $\eta_t \sim U(-0.5, 0.5)$, $v_0 \sim U(-1, 1)$, u_t *and* v_t *are mutually independent,* e_t *are i.i.d. random errors,* $e_t \sim N(0, \sigma_0^2)$, *and* $\{e_t : t \geq 1\}$ *is independent of* $\{(u_t, v_t) : t \geq 1\}$.

Firstly, it is clear that Assumption 6.6.1 holds. See, for example, Lemma 3.1 of Masry and Tjøstheim (1997), Tong (1990) and §2.4 of Doukhan (1995). Secondly, using (6.2.12) and applying the property of trigonometric functions, we have

$$E\{x_t^2 \sin(i\pi x_t)\} = 0 \text{ and } E\{\sin(j\pi x_t)\sin(k\pi x_t)\} = 0$$

for all $i \geq 1$ and $j \neq k$. Therefore Assumption 6.6.3 holds. Finally, it follows that there exists a sequence of constants $\{\gamma_j : j \geq 1\}$ such that the even function $g(v) = \delta v^2$ can be approximated by a special form of the Gallant's flexible Fourier form (see Gallant (1981))

$$\{v^2, \sin(\pi v), \ldots, \sin((q-1)\pi v), \ldots\}.$$

Thus, Assumption 6.6.2 holds. We refer the asymptotic property of trigonometric polynomials to Gallant (1981), Chapter IV of Kashin and Saakyan (1989) and Chapter 7 of DeVore and Lorentz (1993). Thus, Assumptions 6.6.1-6.6.4 hold for Example 6.2.1. Also, Assumptions 6.6.1-6.6.4 can be justified for Example 6.2.2 .

For Example 6.2.1, define the approximation of $g_1(x) = \delta x^2$ by

$$g_1^*(x) = x^2 \gamma_{11} + \sum_{j=2}^{q} \sin(\pi(j-1)x)\gamma_{1j},$$

where $x \in [-1, 1]$, $Z_x(x_t) = \{x_t^2, \sin(\pi x_t), \ldots, \sin((q-1)\pi x_t)\}^T$, $q = 2[T^{1/5}]$, and $\gamma_x = (\gamma_{11}, \ldots, \gamma_{1q})^T$.

For Example 6.2.2, define the approximation of $g_2(v) = 0.75v^2$ by

$$g_2^*(v) = v^2\gamma_{21} + \sum_{l=2}^{q} \sin(\pi(l-1)v)\gamma_{2l},$$

where $v \in [-1, 1]$, $Z_v(v_t) = \{v_t^2, \sin(\pi v_t), \ldots, \sin(\pi(q-1)v_t)\}^T$, $q = 2[T^{1/5}]$, and $\gamma_v = (\gamma_{21}, \ldots, \gamma_{2q})^T$.

For Example 6.2.1, compute

$$L_{1T} = (2q)^{-1/2}\sigma_0^{-2}(\widehat{\gamma}_x^T Z_x^T Z_x \widehat{\gamma}_x - q\sigma_0^2),$$

where $\widehat{\gamma}_x = (Z_x^T Z_x)^{-1} Z_x^T \{\mathcal{F} - U_y(\widehat{U}_y^T \widehat{U}_y)^{-1} \widehat{U}_y^T\} Y$ with $Y = (y_1, \ldots, y_T)^T$ and $U_y = (y_0, y_1, \ldots, y_{T-1})^T$, $Z_x = \{Z_x(x_1), \ldots, Z_x(x_T)\}^T$, $P_x = Z_x(Z_x^T Z_x)^{-1} Z_x^T$, and $\widehat{U}_y = (\mathcal{F} - P_x)U_y$.

For Example 6.2.2, compute

$$L_{2T} = 2^{-1/2}\sigma_0^{-2}(\widehat{\beta}_u^T U_u^T U_u \widehat{\beta}_u - \sigma_0^2),$$

where $\widehat{\beta}_u = (U_u^T U_u)^{-1} U_u^T \{\mathcal{F} - Z_v(\widehat{Z}_v^T \widehat{Z}_v)^{-1} \widehat{Z}_v^T\} Y$ with $Y = (y_1, \ldots, y_T)^T$ and $U_u = (u_1, \ldots, u_T)^T$, $Z_v = \{Z_v(v_1), \ldots, Z_v(v_T)\}^T$, $P_u = U_u(U_u^T U_u)^{-1} U_u^T$, and $\widehat{Z}_v = (\mathcal{F} - P_u)Z_v$.

For Examples 6.2.1 and 6.2.2, we need to find L_0^*, an approximation to the 95-th percentile of L_T. Using the same arguments as in the discussion of Buckley and Eagleson (1988), we can show that a reasonable approximation to the 95th percentile is

$$\frac{\mathcal{X}_{\tau,0.05}^2 - T}{\sqrt{2T}},$$

where $\mathcal{X}_{\tau,0.05}^2$ is the 95th percentile of the chi-squared distribution with τ degrees of freedom. For this example, the critical values L_0^* at $\alpha = 0.05$ were equal to 1.77, 1.74, and 1.72 for T equal to 30, 60, and 100 respectively.

The simulation results below were performed 1200 times and the rejection rates are tabulated in Tables 6.1 and 6.2 below.

Both Tables 6.1 and 6.2 support Theorem 6.2.1. Tables 6.1 and 6.2 also show that the rejection rates seem relatively insensitive to the choice of q, but are sensitive to the values of δ, ϕ and σ_0. The power increased as δ or ϕ increased

TABLE 6.1. Rejection rates for example 6.2.1

T	q	σ_0	$\delta = 0$	$\delta = 0.1$	$\delta = 0.3$	$\delta = 0.6$
30	3	0.1	0.083	0.158	0.55	0.966
60	4	0.1	0.025	0.175	0.766	1.000
100	5	0.1	0.033	0.166	0.941	1.000
30	3	0.2	0.091	0.1	0.191	0.55
60	4	0.2	0.025	0.066	0.291	0.791
100	5	0.2	0.041	0.075	0.35	0.941
30	3	0.25	0.1	0.1	0.183	0.408
60	4	0.25	0.025	0.05	0.2	0.591
100	5	0.25	0.041	0.066	0.233	0.825

TABLE 6.2. Rejection rates for example 6.2.2

T	q	σ_0	$\phi = 0$	$\phi = 0.05$	$\phi = 0.15$	$\phi = 0.25$
30	3	0.1	0.05	0.158	0.733	1.000
60	4	0.1	0.083	0.233	0.941	1.000
100	5	0.1	0.05	0.391	1.000	1.000
30	3	0.2	0.05	0.083	0.233	0.566
60	4	0.2	0.083	0.108	0.491	0.833
100	5	0.2	0.041	0.125	0.716	0.975
30	3	0.3	0.05	0.058	0.158	0.308
60	4	0.3	0.075	0.116	0.233	0.541
100	5	0.3	0.05	0.083	0.391	0.8

while the power decreased as σ_0 increased. Similarly, we compute the rejection rates for the case where both the distributions of e_t and y_0 in Example 6.2.1 are replaced by $U(-0.5, 0.5)$ and $U(-1, 1)$ respectively. Our simulation results show that the performance of L_T under the normal errors is better than that under the uniform errors.

Example 6.2.3 *In this example, we consider the Canadian lynx data. This data set is the annual record of the number of Canadian lynx trapped in the MacKenzie River district of North-West Canada for the years 1821 to 1934. Tong (1977) fitted an eleven th-order linear Gaussian autoregressive model to* $y_t = \log_{10}$*(number of lynx trapped in the year* $(1820 + t)) - 2.9036$ *for* $t = 1, 2, ..., 114$ *(*$T = 114$*), where the average of* $\log_{10}$*(trapped lynx) is 2.9036.*

In the following, we choose y_{n+1} and y_n as the candidates of the regressors and apply Theorem 6.2.1 to test whether the real data set should be fitted by the second-order linear autoregressive model of the form

$$y_{n+2} = \beta_1 y_{n+1} + \beta_2 y_n + e_{1n}, \quad 1 \leq n \leq T \tag{6.2.13}$$

or the second-order `additive autoregressive model` of the form

$$y_{n+2} = \beta_3 y_{n+1} + g(y_n) + e_{2n}, \quad 1 \leq n \leq T,$$

where β_1, β_2 and β_3 are unknown parameters, g is an unknown function, and e_{1n} and e_{2n} are assumed to be i.i.d. random errors with mean zero and finite variance.

For Example 6.2.3, we choose the series functions $z_1(v) = v$ and $\{z_j(v) = \cos((j-1)\pi v) : 2 \leq j \leq q\}$. Our previous research (see Gao, Tong and Wolff (1998a)) on selecting the `truncation parameter` q suggests using $q = 2$ for this example. Similar to Example 6.2.1, the critical value at $\alpha = 0.05$ for L_{1T} with $T = 114$ was 0.6214. With σ_0^2 in (6.2.9) replaced by its estimator $\widehat{\sigma}_0^2 = 0.0419$, the value of L_{1T} was 3.121. Thus the linear model (6.2.13) does not seem appropriate for the lynx data. This is the same as the conclusion reached by Wong and Kohn (1996) through a Bayesian approach.

6.3 Optimum Linear Subset Selection

In Section 6.2, we discussed model (6.2.1). In applications, we need to determine which subset of X_t should become the U_t before using the model (6.2.1) to fit a given set of data. In this section, we construct a CV criterion to select the U_t. In the meantime, we apply this CV criterion to estimate semiparametric regression functions and to model nonlinear time series data. Additionally, we illustrate the consistent CV criterion by simulated and real examples.

6.3.1 A Consistent CV Criterion

Let (Y_t, X_t) be $(r+1)$-dimensional strictly stationary processes with $X_t = (X_{t1}, \ldots, X_{tr})^T$ and $r = p + d$. We write

$$Y_t = m(X_t) + e_t,$$

where $m(x) = E(Y_t|X_t = x)$ and $e_t = Y_t - E(Y_t|X_t)$. For any $A \subset \mathcal{A} \equiv \{1, 2, \ldots, r\}$, we partition X_t into two subvectors U_{tA} and V_{tA}, where U_{tA} consists of $\{X_{ti}, i \in A\}$ and V_{tA} consists of $\{X_{ti}, i \in \mathcal{A} - A\}$. We use $p = \#A$ to denote the cardinality of A and $d = r - p$. We call a d-dimensional function $\phi(x_1, \ldots, x_d)$ completely nonlinear if for any $1 \le i \le d$, ϕ is a nonlinear function of x_i with all other x's fixed.

Before proposing our consistency criterion, we need to make the following assumption.

Assumption 6.3.1 *Suppose that the true unknown regression function is*

$$m(X_t) = U_{tA_0}^T \beta_{A_0} + g_{A_0}(V_{tA_0})$$

for some $A_0 \subset \mathcal{A}$ with $\#A_0 \ge 1$, where β_{A_0} is a constant vector and g_{A_0} is a non-stochastic and completely nonlinear function.

Following Assumption 6.3.1, we have

$$g_{A_0}(v) = g_{1A_0}(v) - g_{2A_0}(v)^T \beta_{A_0},$$

where $g_{1A_0}(v) = E(Y_t|V_{tA_0} = v)$ and $g_{2A_0}(v) = E(U_{tA_0}|V_{tA_0} = v)$.

First, for any given β and $A \subset \mathcal{A}$, we define the following leave-one-out estimators by

$$\begin{aligned}
\widehat{g}_{1t}(V_{tA}, h) &= \sum_{s=1, s \neq t}^{T} W_{sA}(V_{tA}, h) Y_s, \\
\widehat{g}_{2t}(V_{tA}, h) &= \sum_{s=1, s \neq t}^{T} W_{sA}(V_{tA}, h) U_{sA},
\end{aligned}$$

and

$$\widehat{g}_t(V_{tA},\beta) = \widehat{g}_{1t}(V_{tA},h) - \widehat{g}_{2t}(V_{tA},h)^T\beta, \tag{6.3.1}$$

where

$$W_{sA}(V_{tA},h) = K_d\Big(\frac{V_{tA}-V_{sA}}{h}\Big)/\Big\{\sum_{l=1,l\neq t}^{T} K_d\Big(\frac{V_{tA}-V_{lA}}{h}\Big)\Big\},$$

in which K_d is a multivariate kernel function and h is a bandwidth parameter.

Then, we define the kernel-based least squares (LS) estimator $\widehat{\beta}(h,A)$ by minimizing

$$\sum_{t=1}^{T}\{Y_t - U_{tA}^T\widehat{\beta}(h,A) - \widehat{g}_t(V_{tA},\widehat{\beta}(h,A))\}^2.$$

For any given $A \subset \mathcal{A}$ with $|A| \geq 1$, the LS estimator $\widehat{\beta}(h,A)$ is

$$\widehat{\beta}(h,A) = \{\widetilde{\Sigma}(h,A)\}^{+}\sum_{t=1}^{T}\widetilde{U}_{tA}(h)\{Y_t - \widehat{g}_{1t}(V_{tA},h)\},$$

where $\widetilde{U}_{tA}(h) = U_{tA} - \widehat{g}_{2t}(V_{tA},h)$ and $\widetilde{\Sigma}(h,A) = \sum_{t=1}^{T}\widetilde{U}_{tA}(h)\widetilde{U}_{tA}(h)^T$.

For any given $A \subset \mathcal{A}$, we define the following CV function by

$$CV(h,A) = \frac{1}{T}\sum_{t=1}^{T}\{Y_t - U_{tA}^T\widehat{\beta}(h,A) - \widehat{g}_t(V_{tA},\widehat{\beta}(h,A))\}^2. \tag{6.3.2}$$

Remark 6.3.1 *Analogous to Yao and Tong (1994), we avoid using a weight function by assuming that the density of X_t satisfies Assumption 6.6.5(ii) in Section 6.6.*

Let $\widehat{A}_0$ and $\widehat{h}$ denote the estimators of A_0 and h, respectively, which are obtained by minimizing the CV function $CV(h,A)$ over $h \in H_T$ and $A \subset \mathcal{A}$, where $H_T = H_{Td} = \{h_{\min}(T,d), h_{\max}(T,d)\}$ with $0 < h_{\min}(T,d) < h_{\max}(T,d) < 1$ for all T and $d \geq 1$.

The main result of this section is as follows.

Theorem 6.3.1 *Assume that Assumptions 6.3.1, 6.6.1, and 6.6.4- 6.6.7 listed in Section 6.6 hold. Then*

$$\lim_{T\to\infty} \Pr(\widehat{A_0} = A_0) = 1.$$

Remark 6.3.2 *This theorem shows that if some partially linear model within the context tried is the truth, then the CV function will asymptotically find it. Recently, Chen and Chen (1991) considered using a smoothing spline to approximate the nonparametric component and obtained a similar result for the i.i.d. case. See their Proposition 1.*

The proof of Theorem 6.3.1 is postponed to Section 6.6.

Similar to (6.3.1), we define the following estimators of $g_{1A_0}(\cdot)$, $g_{2A_0}(\cdot)$, and $g_{A_0}(\cdot)$ by

$$\begin{aligned} \widehat{g}_1(v;\widehat{h},\widehat{A}_0) &= \frac{1}{T\widehat{h}^{\widehat{d}_0}}\sum_{s=1}^{T}\frac{K_{\widehat{d}_0}((v-V_{s\widehat{A}_0})/\widehat{h})Y_s}{\widehat{f}(v;\widehat{h},\widehat{A}_0)}, \\ \widehat{g}_2(v;\widehat{h},\widehat{A}_0) &= \frac{1}{T\widehat{h}^{\widehat{d}_0}}\sum_{s=1}^{T}\frac{K_{\widehat{d}_0}((v-V_{s\widehat{A}_0})/\widehat{h})U_{s\widehat{A}_0}}{\widehat{f}(v;\widehat{h},\widehat{A}_0)}, \end{aligned} \tag{6.3.3}$$

and

$$\widehat{g}(v;\widehat{h},\widehat{A}_0) = \widehat{g}_1(v;\widehat{h},\widehat{A}_0) - \widehat{g}_2(v;\widehat{h},\widehat{A}_0)^T\widehat{\beta}(\widehat{h},\widehat{A}_0),$$

where $\widehat{d}_0 = r - |\widehat{A}_0|$ and

$$\widehat{f}(v;\widehat{h},\widehat{A}_0) = \frac{1}{T\widehat{h}^{\widehat{d}_0}}\sum_{s=1}^{T}K_{\widehat{d}_0}((v-V_{s\widehat{A}_0})/\widehat{h}). \tag{6.3.4}$$

We now define the estimator of $m(X_t)$ by

$$\widehat{m}(X_t;\widehat{h},\widehat{A}_0) = U_{t\widehat{A}_0}^T\widehat{\beta}(\widehat{h},\widehat{A}_0) + \widehat{g}(V_{t\widehat{A}_0};\widehat{h},\widehat{A}_0).$$

The following result ensures that the prediction error $\widehat{\sigma}^2(\widehat{h},\widehat{A}_0)$ converges to the true variance $\sigma_0^2 = E\{Y_t - m(X_t)\}^2$ in large sample case.

Theorem 6.3.2 *Under the conditions of Theorem 6.3.1, we have as $T \to \infty$*

$$\widehat{\sigma}^2(\widehat{h},\widehat{A}_0) = \frac{1}{T}\sum_{t=1}^{T}\{Y_t - \widehat{m}(X_t;\widehat{h},\widehat{A}_0)\}^2 \longrightarrow^{\mathcal{P}} \sigma_0^2.$$

The proof of Theorem 6.3.2 is mentioned in Section 6.6.

In the following, we briefly mention the application of Theorem 6.3.1 to semiparametric regression and `nonlinear time series` models.

Consider the partially linear model (6.2.1) given by

$$Y_t = U_t^T\beta + g(V_t) + e_t, \tag{6.3.5}$$

For a sequence of independent observations $\{(U_t, V_t) : t \geq 1\}$, model (6.3.5) is a semiparametric regression model. For $Y_t = y_t$, $U_t = (y_{t-c_1}, \ldots, y_{t-c_p})^T$ and $V_t = (y_{t-d_1}, \ldots, y_{t-d_q})$, model (6.3.5) is a **partially linear autoregressive model**. In applications, we need to find the linear regressor U_t before applying the model (6.3.5) to fit real data sets. Obviously, Theorem 6.3.1 can be applied to the two cases.

6.3.2 Simulated and Real Examples

In this section, we apply Theorems 6.3.1 to determine a partially linear ARX model and to fit some real data sets.

Example 6.3.1 *Consider the model given by*

$$y_t = 0.2y_{t-1} + 0.1y_{t-2} + 0.2\sin(\pi x_t) + e_t, t = 2, 3, ..., T, \tag{6.3.6}$$

where $x_t = 0.5x_{t-1} + \varepsilon_t$, e_t and ε_t are mutually independent and identically distributed over uniform $(-0.5, 0.5)$, x_1, y_0 and y_1 are mutually independent and identically distributed over uniform $(-1, 1)$, and both ε_t and e_t are independent of (x_1, y_0, y_1).

In this example, we consider using the following kernel function

$$K_d(u_1, u_2, \ldots, u_d) = \prod_{i=1}^{d} K(u_i),$$

where $d = 1, 2, 3$,

$$K(u) = \begin{cases} (15/16)(1-u^2)^2 & \text{if } |u| \leq 1 \\ 0 & \text{otherwise} \end{cases}$$

and $H_T = [T^{-7/30}, 1.1T^{-1/6}]$.

Select y_{t-1}, y_{t-2} and x_t as the candidates of the regressors. In this example, $\mathcal{A} = \{1, 2, 3\}$. Let A be the linear subset of $\mathcal{A}$. Through computing the CV function given in (6.3.2), we obtain the results listed in Table 6.3.

In Table 6.3, $A = \{1, 2\}$ means that y_{t-1} and y_{t-2} are the linear regressors, $A = \{1, 3\}$ means that y_{t-1} and x_t are the linear regressors, $A = \{2, 3\}$ means that y_{t-2} and x_t are the linear regressors, and $A = \{1, 2, 3\}$ means that (6.3.6) is a linear ARX model.

In Example 6.3.2, we select y_{n+2} as the present observation and both y_{n+1} and y_n as the candidates of the regressors, $n = 1, 2, \ldots, T$.

TABLE 6.3. Frequencies of selected linear subsets in 100 replications for example 6.2.3

Linear subsets	T=51	T=101	T=201
A={1,2}	86	90	95
A={1,3}	7	5	3
A={2,3}	5	4	2
A={1,2,3}	2	1	

Example 6.3.2 *In this example, we consider using Theorem 6.3.1 to fit the sunspots data (Data I) and the Australian blowfly data (Data II). For Data I, first normalize the data X by $X^* = \{X - \text{mean}(X)\}/\{\text{var}(X)\}^{1/2}$ and define y_t = the normalized sunspot number in the year $(1699 + t)$, where $1 \le t \le 289$ $(T = 289)$, $\text{mean}(X)$ denotes the sample mean and $\text{var}(X)$ denotes the sample standard deviation. For Data II, we take a log transformation of the data by defining $y = \log_{10}$(blowfly population) first and define $y_t = \log_{10}$(blowfly population number at time t) for $t = 1, 2, \ldots, 361$ $(T = 361)$.*

In the following, we only consider the case where $\mathcal{A} = \{1, 2\}$ and apply the consistent CV criterion to determine which model among the following possible models (6.3.7)–(6.3.8) should be selected to fit the real data sets,

$$(I) \quad y_{n+2} = \beta_1 y_n + g_2(y_{n+1}) + e_{1n}, \tag{6.3.7}$$

$$(II) \quad y_{n+2} = \beta_2 y_{n+1} + g_1(y_n) + e_{2n}, \tag{6.3.8}$$

where β_1 and β_2 are unknown parameters, g_1 and g_2 are unknown and nonlinear functions, and e_{1n} and e_{2n} are assumed to be i.i.d. random errors with zero mean and finite variance.

Our experience suggests that the choice of the kernel function is much less critical than that of the bandwidth. In the following, we choose the kernel function $K(x) = (2\pi)^{-1/2}\exp(-x^2/2)$ and $h \in H_T = [T^{-7/30}, 1.1T^{-1/6}]$.

Through computing the CV function defined by (6.3.2) for Example 6.3.2, we can obtain the following minimum CV values listed in Table 6.4.

For the two data sets, when selecting y_{n+1} and y_n as the candidates of the regressors, Table 6.4 suggests using the prediction equation

$$\widehat{y}_{n+2} = \widehat{\beta}_2(\widehat{h}_{2C})y_{n+1} + \widetilde{g}_1(y_n), n = 1, 2, \ldots, \tag{6.3.9}$$

TABLE 6.4. The minimum CV values for example 6.3.1

Models	CV–Data I	CV–Data II
I	0.1914147	0.04251451
II	0.1718859	0.03136296

where $\widetilde{g}_1(y_n) = \widehat{g}_2(y_n, \widehat{h}_{2C}) - \widehat{\beta}_2(\widehat{h}_{2C})\widehat{g}_1(y_n, \widehat{h}_{2C})$ appears to be nonlinear,

$$\widehat{g}_i(y_n, h) = \Big\{\sum_{m=1}^{N} K(\frac{y_n - y_m}{h})y_{m+i}\Big\}/\Big\{\sum_{m=1}^{N} K(\frac{y_n - y_m}{h})\Big\}, i = 1, 2,$$

and $\widehat{h}_{2C} = 0.2666303$ and 0.3366639, respectively.

In the following, we consider the case where $\mathcal{A} = \{1, 2, 3\}$ and apply the consistent CV criterion to determine which model among the following possible models (6.3.10)–(6.3.15) should be selected to fit the real data sets given in Examples 6.3.3 and 6.3.4 below,

$$(M1) \quad Y_t = \beta_1 X_{t1} + \beta_2 X_{t2} + g_3(X_{t3}) + e_{1t}, \qquad (6.3.10)$$

$$(M2) \quad Y_t = \beta_3 X_{t1} + g_2(X_{t2}) + \beta_4 X_{t3} + e_{2t}, \qquad (6.3.11)$$

$$(M3) \quad Y_t = g_1(X_{t1}) + \beta_5 X_{t2} + \beta_6 X_{t3} + e_{3t}, \qquad (6.3.12)$$

$$(M4) \quad Y_t = \beta_7 X_{t1} + G_1(X_{1t}) + e_{4t}, \qquad (6.3.13)$$

$$(M5) \quad Y_t = \beta_8 X_{t2} + G_2(X_{2t}) + e_{5t}, \qquad (6.3.14)$$

$$(M6) \quad Y_t = \beta_9 X_{t3} + G_3(X_{3t}) + e_{6t}, \qquad (6.3.15)$$

where $X_{1t} = (X_{t2}, X_{t3})^T$, $X_{2t} = (X_{t1}, X_{t3})^T$, $X_{3t} = (X_{t1}, X_{t2})^T$, β_i are unknown parameters, (g_j, G_j) are unknown and nonlinear functions, and e_{it} are assumed to be i.i.d. random errors with zero mean and finite variance.

Analogously, we can compute the corresponding CV functions with K and h chosen as before for the following examples.

Example 6.3.3 *In this example, we consider the data, given in Table 5.1 of Daniel and Wood (1971), representing 21 successive days of operation of a plant oxidizing ammonia to nitric acid. Factor x_1 is the flow of air to the plant. Factor x_2 is the temperature of the cooling water entering the countercurrent nitric oxide absorption tower. Factor x_3 is the concentration of nitric acid in the absorbing liquid. The response, y, is 10 times the percentage of the ingoing ammonia that is lost as unabsorbed nitric oxides; it is an indirect measure of the yield of nitric*

TABLE 6.5. The minimum CV values for examples 6.3.2 and 6.3.3

Models	CV–Example 6.3.2	CV–Example 6.3.3
(M1)	0.008509526	0.06171114
(M2)	0.005639502	0.05933659
(M3)	0.005863942	0.07886344
(M4)	0.007840915	0.06240574
(M5)	0.007216028	0.08344121
(M6)	0.01063646	0.07372809

acid. From the research of Daniel and Wood (1971), we know that the transformed response $\log_{10}(y)$ *depends nonlinearly on some subset of* (x_1, x_2, x_3). *In the following, we apply the above Theorem 6.3.1 to determine what is the true relationship between* $\log_{10}(y)$ *and* (x_1, x_2, x_3).

Example 6.3.4 *We analyze the transformed Canadian lynx data* $y_t = \log_{10}($*number of lynx trapped in the year* $(1820 + t))$ *for* $1 \le t \le T = 114$. *The research of Yao and Tong (1994) has suggested that the subset* $(y_{t-1}, y_{t-3}, y_{t-6})$ *should be selected as the candidates of the regressors when estimating the relationship between the present observation* y_t *and* $(y_{t-1}, y_{t-2}, \ldots, y_{t-6})$. *In this example, we apply the consistent CV criterion to determine whether* y_t *depends linearly on a subset of* $(y_{t-1}, y_{t-3}, y_{t-6})$.

For Example 6.3.3, let $Y_t = \log_{10}(y_t)$, $X_{t1} = x_{t1}$, $X_{t2} = x_{t2}$, and $X_{t3} = x_{t3}$ for $t = 1, 2, \ldots, T$. For Example 6.3.4, let $Y_t = y_{t+6}$, $X_{t1} = y_{t+5}$, $X_{t2} = y_{t+3}$, and $X_{t3} = y_t$ for $t = 1, 2, \ldots, T - 6$. Through minimizing the corresponding CV functions, we obtain the following minimum CV and the CV-based $\widehat{\beta}(\widehat{h}_C)$ values listed in Tables 6.5 and 6.6, respectively.

For Example 6.3.3, when selecting (6.3.11), (6.3.12) or (6.3.15) to fit the data, Table 6.6 shows that the factor x_3's influence can be negligible since the CV-based coefficients $\widehat{\beta}_4(\widehat{h}_{2C})$, $\widehat{\beta}_6(\widehat{h}_{3C})$ and $\widehat{\beta}_9(\widehat{h}_{6C})$ are relatively smaller than the other coefficients. This conclusion is the same as that of Daniel and Wood (1971), who analyzed the data by using classical linear regression models. Table 6.5 suggests using the following prediction equation

$$\widehat{y}_t = \widehat{\beta}_3(\widehat{h}_{2C})x_{t1} + \widetilde{g}_2(x_{t2}) + \widehat{\beta}_4(\widehat{h}_{2C})x_{t3}, t = 1, 2, \ldots, 21,$$

TABLE 6.6. The CV-based $\widehat{\beta}(\widehat{h}_C)$ values for examples 6.3.2 and 6.3.3

$\widehat{\beta}(\widehat{h}_C)$-Value	Example 6.3.2	Example 6.3.4
$\widehat{\beta}_1(\widehat{h}_{1C})$	0.01590177	0.9824141
$\widehat{\beta}_2(\widehat{h}_{1C})$	0.0273703	-0.4658888
$\widehat{\beta}_3(\widehat{h}_{2C})$	0.01744958	0.9403801
$\widehat{\beta}_4(\widehat{h}_{2C})$	-0.003478075	0.03612305
$\widehat{\beta}_5(\widehat{h}_{3C})$	0.04129525	-0.4313919
$\widehat{\beta}_6(\widehat{h}_{3C})$	-0.001771901	0.02788928
$\widehat{\beta}_7(\widehat{h}_{4C})$	0.01964107	0.9376097
$\widehat{\beta}_8(\widehat{h}_{5C})$	0.04329919	-0.423475
$\widehat{\beta}_9(\widehat{h}_{6C})$	-0.003640449	0.03451025

where $\widetilde{g}_2(x_{t2}) = \widehat{g}_2(x_{t2}, \widehat{h}_{2C}) - \{\widehat{\beta}_3(\widehat{h}_{2C}), \widehat{\beta}_4(\widehat{h}_{2C})\}^T \widehat{g}_1(x_{t2}, \widehat{h}_{2C})$ appears to be nonlinear,

$$\widehat{g}_i(x_{t2}, h) = \Big\{\sum_{s=1}^{21} K(\frac{x_{t2} - x_{s2}}{h}) Z_{is}\Big\} \Big/ \Big\{\sum_{s=1}^{21} K(\frac{x_{t2} - x_{s2}}{h})\Big\}, \quad i = 1, 2,$$

$Z_{2s} = Y_s$, $Z_{1s} = (x_{s1}, x_{s2})^T$, and $\widehat{h}_{2C} = 0.6621739$.

For the Canadian lynx data, when selecting y_{n+5}, y_{n+3}, and y_n as the candidates of the regressors, Tables 6.5 and 6.6 suggest using the following prediction equation

$$\widehat{y}_{n+6} = \widehat{\beta}_3(\widehat{h}_{2C}) y_{n+5} + \widetilde{g}_2(y_{n+3}) + \widehat{\beta}_4(\widehat{h}_{2C}) y_n, n = 1, 2, \ldots, 108, \qquad (6.3.16)$$

where $\widetilde{g}_2(y_{n+3}) = \widehat{g}_2(y_{n+3}, \widehat{h}_{2C}) - \{\widehat{\beta}_3(\widehat{h}_{2C}), \widehat{\beta}_4(\widehat{h}_{2C})\}^T \widehat{g}_1(y_{n+3}, \widehat{h}_{2C})$ appears to be nonlinear,

$$\widehat{g}_i(y_{n+3}, h) = \Big\{\sum_{m=1}^{108} K(\frac{y_{m+3} - y_{n+3}}{h}) z_{im}\Big\} \Big/ \Big\{\sum_{m=1}^{108} K(\frac{y_{m+3} - y_{n+3}}{h})\Big\}, \quad i = 1, 2,$$

$z_{2m} = y_{m+6}$, $z_{1m} = (y_{m+5}, y_m)^T$, and $\widehat{h}_{2C} = 0.3312498$.

The research by Tong (1990) and Chen and Tsay (1993) has suggested that the fully nonparametric autoregressive model of the form $y_t = \widetilde{g}(y_{t-1}, \ldots, y_{t-r}) + e_t$ is easier to understand than the threshold autoregressive approach proposed by Tong (1977). It follows from equations (6.3.9) and (6.3.16) that for the Canadian lynx data, the sunspots data and the Australian blowfly data, the above `partially linear autoregressive model` of the form (6.3.5) is more appropriate than the fully `nonparametric autoregressive model`.

6.4 Optimum Bandwidth Selection

6.4.1 Asymptotic Theory

As mentioned in Tong (1990), some nonlinear phenomena cannot be fitted by linear ARMA models and therefore the fully nonparametric autoregressive function approach is recommended to use in practice. But in some cases, the fully nonparametric autoregressive function approach will lose too much information on the relationship between $\{y_t\}$ and $\{y_{t-i}, i \geq 1\}$ and neglect some existing linear dependencies among them. A reasonable approach to modelling the nonlinear time series data is to use the partially linear additive autoregressive model

$$y_t = \sum_{i=1}^{p} \beta_i y_{t-c_i} + \sum_{j=1}^{q} g_j(y_{t-d_j}) + e_t,$$

where $t > \max(c_p, d_q)$, $1 \leq c_1 < \ldots < c_p \leq r$, $1 \leq d_1 < \ldots < d_q \leq r$, and $c_i \neq d_j$ for all $1 \leq i \leq p$ and $1 \leq j \leq q$.

There are a number of practical motivations for the study of the above model. These include the research of population biology model and the Mackey-Glass system (Glass and Mackey, 1988). Recently, Nychka, Elliner, Gallant and McCaffrey (1992) suggested studying the model

$$y_t = a y_{t-1} + b \frac{y_{t-d}}{1 + y_{t-d}^k} + e_t.$$

For $k = 10$ this is a discretized version of the Mackey-Glass delay differential equation, originally developed to model the production and loss of white blood cells. It can also be interpreted as a model for population dynamics. If $0 < a < 1$ and $b > 0$ and if $\{y_t\}$ denotes the number of adults, then a is the survival rate of adults and d is the time delay between birth and maturation. The $\{b y_{t-d}(1 + y_{t-d}^k)^{-1}\}$ accounts for the recruitment of new adults due to births d years in the past, which is non-linear because of decreased fecundity at higher population levels. In addition, the development in partially linear (semiparametric) regression has established a solid foundation for partially linear time series analysis.

For simplicity, we only consider a partially linear autoregressive model of the form

$$y_t = \beta y_{t-1} + g(y_{t-2}) + e_t, \quad t = 3, 4, \ldots, T \tag{6.4.1}$$

where β is an unknown parameter-of-interest, g is an unknown function over $R^1 = (-\infty, \infty)$, $\{e_t : t \geq 3\}$ is a sequence of i.i.d. random errors with $Ee_1 = 0$ and $Ee_1^2 = \sigma^2 < \infty$, and $\{e_t : t \geq 3\}$ is independent of (y_1, y_2).

For identifiability, we assume that the (β, g) satisfies

$$E\{y_t - \beta y_{t-1} - g(y_{t-2})\}^2 = \min_{(\alpha,f)} E\{y_t - \alpha y_{t-1} - f(y_{t-2})\}^2.$$

It follows from (6.4.1) that

$$\begin{aligned} g(y_{t-2}) &= E\{(y_t - \beta y_{t-1})|y_{t-2}\} \\ &= E(y_t|y_{t-2}) - \beta E(y_{t-1}|y_{t-2}) = g_1(y_{t-2}) - \beta g_2(y_{t-2}). \end{aligned}$$

The natural estimates of g_i $(i = 1, 2)$ and g can be defined by

$$\begin{aligned} \widehat{g}_{1,h}(y_{t-2}) &= \sum_{s=3}^{T} W_{s,h}(y_{t-2}) y_s, \\ \widehat{g}_{2,h}(y_{t-2}) &= \sum_{s=3}^{T} W_{s,h}(y_{t-2}) y_{s-1}, \end{aligned}$$

and

$$\widehat{g}_h(y_{t-2}) = \widehat{g}_{1,h}(y_{t-2}) - \beta \widehat{g}_{2,h}(y_{t-2}),$$

where $\{W_{s,h}(\cdot)\}$ is a probability weight function depending on $y_1, y_2, ..., y_{T-2}$ and the number T of observations.

Based on the model $y_t = \beta y_{t-1} + \widehat{g}_h(y_{t-2}) + e_t$, the kernel-weighted least squares (LS) estimator $\widehat{\beta}(h)$ of β can be defined by minimizing

$$\sum_{t=3}^{T} \{y_t - \beta y_{t-1} - \widehat{g}_h(y_{t-2})\}^2.$$

We now obtain

$$\widehat{\beta}(h) - \beta = \Big(\sum_{t=3}^{T} u_t^2\Big)^{-1} \Big\{\sum_{t=3}^{T} u_t e_t + \sum_{t=3}^{T} u_t \bar{g}_h(y_{t-2})\Big\}, \tag{6.4.2}$$

where $u_t = y_{t-1} - \widehat{g}_{2,h}(y_{t-2})$ and $\bar{g}_h(y_{t-2}) = g(y_{t-2}) - \widehat{g}_h(y_{t-2})$.

In this section, we only consider the case where the $\{W_{s,h}(\cdot)\}$ is a kernel weight function

$$W_{s,h}(x) = K_h(x - y_{s-2}) \Big/ \sum_{t=3}^{T} K_h(x - y_{t-2}),$$

where $K_h(\cdot) = h^{-1}K(\cdot/h)$, $K : R \to R$ is a kernel function satisfying Assumption 6.6.8 below and

$$h = h_T \in H_T = [a_1 T^{-1/5-c_1}, b_1 T^{-1/5+c_1}],$$

in which the absolute constants a_1, b_1 and c_1 satisfy $0 < a_1 < b_1 < \infty$ and $0 < c_1 < 1/20$.

A mathematical measurement of the proposed estimators can be obtained by considering the average squared error (ASE)

$$D(h) = \frac{1}{T-2}\sum_{t=3}^{T}[\{\widehat{\beta}(h)y_{t-1} + \widehat{g}_h^*(y_{t-2})\} - \{\beta y_{t-1} + g(y_{t-2})\}]^2 w(y_{t-2}),$$

where $\widehat{g}_h^*(\cdot) = \widehat{g}_{1,h}(\cdot) - \widehat{\beta}(h)\widehat{g}_{2,h}(\cdot)$ and w is a weight function. It follows from the ASE (see Lemma 6.6.7 (i) below) that the theoretically optimum bandwidth is proportional to $n^{-1/5}$. Unfortunately, this optimization procedure has the drawback that it involves functionals of the underlying distribution. In this section, we will propose a practical selection procedure and then discuss adaptive estimates.

We now have the main results of this section.

Theorem 6.4.1 *Assume that Assumption 6.6.8 holds. Let $Ee_1 = 0$ and $Ee_1^2 = \sigma^2 < \infty$. Then the following holds uniformly over $h \in H_T$*

$$\sqrt{T}\{\widehat{\beta}(h) - \beta\} \longrightarrow^{\mathcal{L}} N(0, \sigma^2\sigma_2^{-2}),$$

where $\sigma_2^2 = E\{y_{t-1} - E(y_{t-1}|y_{t-2})\}^2$.

Theorem 6.4.2 *Assume that Assumption 6.6.8 holds. Let $Ee_1 = 0$ and $Ee_1^4 < \infty$. Then the following holds uniformly over $h \in H_T$*

$$\sqrt{T}\{\widehat{\sigma}(h)^2 - \sigma^2\} \longrightarrow^{\mathcal{L}} N(0, Var(e_1^2)),$$

where $\widehat{\sigma}(h)^2 = 1/T\sum_{t=3}^{T}\{\widehat{y}_t - \widehat{\beta}(h)\widehat{y}_{t-1}\}^2$, $\widehat{y}_{t-1} = y_{t-1} - \widehat{g}_{2,h}(y_{t-2})$, and $\widehat{y}_t = y_t - \widehat{g}_{1,h}(y_{t-2})$.

Remark 6.4.1 *(i) Theorem 6.4.1 shows that the kernel-based estimator of β is asymptotically normal with the smallest possible asymptotic variance (Chen (1988)).*

(ii) Theorems 6.4.1 and 6.4.2 only consider the case where $\{e_t\}$ is a sequence of i.i.d. random errors with $Ee_t = 0$ and $Ee_t^2 = \sigma^2 < \infty$. As a matter of fact,

both Theorems 6.4.1 and 6.4.2 can be modified to the case where $Ee_t = 0$ *and* $Ee_t^2 = f(y_{t-1})$ *with some unknown function* $f > 0$. *For this case, we need to construct an estimator for* f. *For example,*

$$\widehat{f}_T(y) = \sum_{t=3}^{T} \widetilde{W}_{T,t}(y)\{y_t - \widehat{\beta}(h)y_{t-1} - \widehat{g}_h^*(y_{t-2})\}^2,$$

where $\{\widetilde{W}_{T,t}(y)\}$ *is a kernel weight function,* $\widehat{\beta}(h)$ *is as defined in (6.4.2) and* $\widehat{g}_h^*(y_{t-2}) = \widehat{g}_{1,h}(y_{t-2}) - \widehat{\beta}(h)\widehat{g}_{2,h}(y_{t-2})$. *Similar to the proof of Theorem 6.4.1, we can show that* $\widehat{f}_T(y)$ *is a consistent estimator of* f. *Then, we define a weighted LS estimator* $\bar{\beta}(h)$ *of* β *by minimizing*

$$\sum_{t=3}^{T} \widehat{f}_T(y_{t-1})^{-1}\{y_t - \beta y_{t-1} - \widehat{g}_h(y_{t-2})\}^2,$$

where $\widehat{g}_h(y_{t-2})$ *is as defined in (6.4.2).*

Some additional conditions on f *are required to establish the corresponding results of Theorems 6.4.1 and 6.4.2.*

Remark 6.4.2 *A generalization of model (6.4.1) is*

$$y_t = \sum_{s=1}^{p} \beta_s y_{t-s} + g(y_{t-p-1}) + e_t = x_t^T\beta + g(y_{t-p-1}) + e_t, \; t \geq p+2, \quad (6.4.3)$$

where $x_t = (y_{t-1}, \ldots, y_{t-p})^T$, $\beta = (\beta_1, \cdots, \beta_p)^T$ *is a vector of unknown parameters, and the* g *and* $\{e_t\}$ *are as defined in (6.4.1). For this case, we need to modify the above equations (see §4.2 of Chapter III of Györfi, Härdle, Sarda and Vieu (1989)) and the kernel-weighted LS estimator of* β *can be defined as*

$$\widehat{\beta} = \Big(\sum_{t=p+2}^{T} U_t U_t^T\Big)^+ \sum_{t=p+2}^{T} U_t V_t,$$

where

$$\begin{aligned} V_t &= y_t - \widehat{g}_0(y_{t-p-1}), \; U_t = x_t - G(y_{t-p-1}), \\ G(\cdot) &= \{\widehat{g}_1(\cdot), \ldots, \widehat{g}_p(\cdot)\}^T, \; \widehat{g}_i(\cdot) = \sum_{s=p+2}^{T} W_{s,h}(\cdot) y_{s-i} \end{aligned}$$

for $i = 0, 1, \ldots, p$, *in which*

$$W_{s,h}(\cdot) = K_h(\cdot - y_{s-p-1}) \Big/ \Big\{\sum_{l=p+2}^{T} K_h(\cdot - y_{l-p-1})\Big\}.$$

Under similar conditions, the corresponding results of Theorems 6.4.1 and 6.4.2 can be established.

Remark 6.4.3 *Theorems 6.4.1 and 6.4.2 only establish the asymptotic results for the partially linear model (6.4.1). In practice, we need to determine whether model (6.4.1) is more appropriate than*

$$y_t = f_1(y_{t-1}) + \alpha y_{t-2} + \epsilon_t, \tag{6.4.4}$$

where f_1 is an unknown function over R^1, α is an unknown parameter and $\{\epsilon_t\}$ is a sequence of i.i.d. random errors with mean zero and finite variance. In this case, we need to modify the above estimation equations and to estimate the regression function of $E(y_{t-2}|y_{t-1})$, which has been discussed in §5.2.4 of Tong (1990) and Robinson (1983). They both have discussed the estimators of $E(y_t|y_{t\pm j})$ for $j \geq 1$. Therefore, similar results for (6.4.4) can also be obtained. Section 6.4.2 below provides estimation procedures for both (6.4.1) and (6.4.4).

In the following section, we apply a cross-validation (CV) criterion to construct an asymptotically optimal data-driven bandwidth and adaptive data-driven estimates.

Let us define $N = T - 2$,

$$\begin{aligned} D(h) &= \frac{1}{T-2}\sum_{t=3}^{T}[\{\widehat{\beta}(h)y_{t-1} + \widehat{g}_h^*(y_{t-2})\} - \{\beta y_{t-1} + g(y_{t-2})\}]^2 w(y_{t-2}) \\ &= \frac{1}{N}\sum_{n=1}^{N}[\{\widehat{\beta}(h)y_{n+1} + \widehat{g}_h^*(y_n)\} - \{\beta y_{n+1} + g(y_n)\}]^2 w(y_n), \end{aligned} \tag{6.4.5}$$

and

$$CV(h) = \frac{1}{N}\sum_{n=1}^{N}[y_{n+2} - \{\widetilde{\beta}(h)y_{n+1} + \widehat{g}_{1,n}(y_n) - \widetilde{\beta}(h)\widehat{g}_{2,n}(y_n)\}]^2 w(y_n), \tag{6.4.6}$$

where $\widetilde{\beta}(h)$ is as defined in (6.4.2) with $\widehat{g}_h(\cdot)$ replaced by $\widehat{g}_{h,n}(\cdot) = \widehat{g}_{1,n}(\cdot) - \beta\widehat{g}_{2,n}(\cdot)$, in which

$$\widehat{g}_{i,n}(\cdot) = \widehat{g}_{i,n}(\cdot, h) = \frac{1}{N-1}\sum_{m \neq n} K_h(\cdot - y_m)y_{m+3-i}/\widehat{f}_{h,n}(\cdot)$$

and

$$\widehat{f}_{h,n}(\cdot) = \frac{1}{N-1}\sum_{m \neq n} K_h(\cdot - y_m). \tag{6.4.7}$$

Definition 6.4.1 *A data-driven bandwidth* $\widehat{h}$ *is* asymptotically optimal *if*

$$\frac{D(\widehat{h})}{\inf_{h \in H_T} D(h)} \to_p 1.$$

CROSS–VALIDATION (CV): Select h, denoted by $\widehat{h}_C$, that achieves

$$CV(\widehat{h}_C) = \inf_{h \in H_{N+2}} CV(h). \tag{6.4.8}$$

Theorem 6.4.3 *Assume that the conditions of Theorem 6.4.1 hold. Then the data-driven bandwidth* $\widehat{h}_C$ *is asymptotically optimal.*

Theorem 6.4.4 *Assume that the conditions of Theorem 6.4.1 hold. Then under the null hypothesis* $H_0 : \beta = 0$

$$\widehat{F}_1(\widehat{h}_C) = T\{\hat{\beta}(\widehat{h}_C)\}^2 \sigma_2^2 \sigma^{-2} \longrightarrow^{\mathcal{L}} \chi^2(1), \tag{6.4.9}$$

as $T \to \infty$*. Furthermore, under* $H_1 : \beta \neq 0$*, we have* $\widehat{F}(\widehat{h}_C) \to \infty$ *as* $T \to \infty$*.*

Theorem 6.4.5 *Assume that the conditions of Theorem 6.4.1 hold. Then under the null hypothesis* $H_0' : \sigma^2 = \sigma_0^2$

$$\widehat{F}_2(\widehat{h}_C) = T\{\widehat{\sigma}(\widehat{h}_C)^2 - \sigma_0^2\}^2 \{Var(e_1^2)\}^{-1} \to \chi^2(1), \tag{6.4.10}$$

as $T \to \infty$*. Furthermore, under* $H_1' : \sigma^2 \neq \sigma_0^2$*, we have* $\widehat{F}(\widehat{h}_C) \to \infty$ *as* $T \to \infty$*.*

Remark 6.4.4 *Theorems 6.4.3-6.4.5 show that the optimum data-driven bandwidth* $\widehat{h}_C$ *is asymptotically optimal and the conclusions of Theorems 6.4.1 and 6.4.2 remain unchanged with* h *replaced by* $\widehat{h}_C$*. It follows from Theorems 6.4.4 and 6.4.5 that when* h *is proportional to* $n^{-1/5}$*, both* $\widehat{\beta}$ *and* $\widehat{\sigma}^2$ *are* $\sqrt{n}$*–consistent. In addition, it follows from Lemma 6.6.7(i) that the nonparametric estimate* $\widehat{g}_h^*$ *is of* $n^{-4/5}$ *rate of mean squared error (MSE) when* h *is proportional to* $n^{-1/5}$*.*

6.4.2 Computational Aspects

In this subsection, we demonstrate how well the above estimation procedure works numerically and practically.

Example 6.4.1 *Consider model (6.4.1) given by*

$$y_t = \beta y_{t-1} + g(y_{t-2}) + e_t, \ t = 3, 4, ..., T, \tag{6.4.11}$$

where $\{e_t : t \geq 3\}$ *is independent and uniformly distributed over* $(-0.5, 0.5)$, y_1 *and* y_2 *are mutually independent and identically distributed over* $(-1, 1)$, (y_1, y_2) *is independent of* $\{e_t : t \geq 3\}$, *and* (β, g) *is chosen from one of the following models.*

Model 1. $\beta = 0.25$ and $g(y) = y/2(1 + y^2)^{-1}$.

Model 2. $\beta = 0.25$ and $g(y) = 1/4 \sin(\pi y)$.

In this section, we conduct a small sample study for the two models.

Choose the quartic kernel function

$$K(u) = \begin{cases} (15/16)(1 - u^2)^2 & \text{if } |u| \leq 1 \\ 0 & \text{otherwise} \end{cases}$$

and the weight function

$$w(x) = \begin{cases} 1 & \text{if } |x| \leq 1 \\ 0 & \text{otherwise} \end{cases}$$

First, because of the form of g, the fact that the process $\{y_t\}$ is strictly stationary follows from §2.4 of Tjøstheim (1994) (also Theorem 3.1 of An and Huang (1996)). Second, by using Lemma 3.4.4 and Theorem 3.4.10 of Györfi, Härdle, Sarda and Vieu (1989). (also Theorem 7 of §2.4 of Doukhan (1995)), we obtain that the $\{y_t\}$ is β-mixing and therefore α-mixing. Thus, Assumption 6.6.1 holds. Third, it follows from the definition of K and w that Assumption 6.6.8 holds.

(i). Based on the simulated data set $\{y_t : 1 \leq t \leq T\}$, compute the following estimators

$$\begin{aligned}
\widehat{g}_{1,h}(y_n) &= \Big\{\sum_{m=1}^{N} K_h(y_n - y_m) y_{m+2}\Big\} \Big/ \Big\{\sum_{m=1}^{N} K_h(y_n - y_m)\Big\}, \\
\widehat{g}_{2,h}(y_n) &= \Big\{\sum_{m=1}^{N} K_h(y_n - y_m) y_{m+1}\Big\} \Big/ \Big\{\sum_{m=1}^{N} K_h(y_n - y_m)\Big\}, \\
\widehat{g}_h(y_n) &= \widehat{g}_{1,h}(y_n) - \frac{1}{4}\widehat{g}_{2,h}(y_n), \\
\widehat{g}_{1,n}(y_n) &= \Big\{\sum_{m \neq n}^{N} K_h(y_n - y_m) y_{m+2}\Big\} \Big/ \Big\{\sum_{m \neq n}^{N} K_h(y_n - y_m)\Big\}, \\
\widehat{g}_{2,n}(y_n) &= \Big\{\sum_{m \neq n}^{N} K_h(y_n - y_m) y_{m+1}\Big\} \Big/ \Big\{\sum_{m \neq n}^{N} K_h(y_n - y_m)\Big\},
\end{aligned}$$

and

$$\widehat{g}_n(y_n) = \widehat{g}_{1,n}(y_n) - \frac{1}{4}\widehat{g}_{2,n}(y_n), \qquad (6.4.12)$$

where $K_h(\cdot) = h^{-1}K(\cdot/h)$, $h \in H_{N+2} = [(N+2)^{-7/30}, 1.1(N+2)^{-1/6}]$, and $1 \leq n \leq N$.

(ii). Compute the LS estimates $\widehat{\beta}(h)$ of (6.4.2) and $\widetilde{\beta}(h)$ of (6.4.6)

$$\widehat{\beta}(h) - \beta = \Big(\sum_{n=1}^{N} u_{n+2}^2\Big)^{-1}\Big\{\sum_{n=1}^{N} u_{n+2}e_{n+2} + \sum_{n=1}^{N} u_{n+2}\bar{g}_h(y_n)\Big\} \tag{6.4.13}$$

and

$$\widetilde{\beta}(h) - \beta = \Big(\sum_{n=1}^{N} v_{n+2}^2\Big)^{-1}\Big\{\sum_{n=1}^{N} v_{n+2}e_{n+2} + \sum_{n=1}^{N} v_{n+2}\bar{g}_n(y_n)\Big\} \tag{6.4.14}$$

where $u_{n+2} = y_{n+1} - \widehat{g}_{2,h}(y_n)$, $\bar{g}_h(y_n) = g(y_n) - \widehat{g}_h(y_n)$, $v_{n+2} = y_{n+1} - \widehat{g}_{2,n}(y_n)$, and $\bar{g}_n(y_n) = g(y_n) - \widehat{g}_n(y_n)$.

(iii). Compute

$$\begin{aligned} D(h) &= \frac{1}{N}\sum_{n=1}^{N}\{\widehat{m}_h(z_n) - m(z_n)\}^2 \\ &= \frac{1}{N}\sum_{n=1}^{N}[\{\widehat{\beta}(h)u_{n+2} + \widehat{g}_{1,h}(y_n)\} - \{\beta y_{n+1} + g(y_n)\}]^2 \\ &= \frac{1}{N}\sum_{n=1}^{N}\{\widehat{\beta}(h)u_{n+2} + \widehat{g}_{1,h}(y_n)\}^2 \\ &\quad -\frac{2}{N}\sum_{n=1}^{N}\{\widehat{\beta}(h)u_{n+2} + \widehat{g}_{1,h}(y_n)\}\{\beta y_{n+1} + g(y_n)\} \\ &\quad +\frac{1}{N}\sum_{n=1}^{N}\{\beta y_{n+1} + g(y_n)\}^2 \\ &\equiv D_{1h} + D_{2h} + D_{3h}, \end{aligned} \tag{6.4.15}$$

where the symbol "$\equiv$" indicates that the terms of the left-hand side can be represented by those of the right-hand side correspondingly.

Thus, D_{1h} and D_{2h} can be computed from (6.4.6)–(6.4.15). D_{3h} is independent of h. Therefore the problem of minimizing $D(h)$ over H_{N+2} is the same as that of minimizing $D_{1h} + D_{2h}$. That is

$$\widehat{h}_D = \arg\min_{h\in H_{N+2}} (D_{1h} + D_{2h}). \tag{6.4.16}$$

(iv). Compute

$$\begin{aligned} CV(h) &= \frac{1}{N}\sum_{n=1}^{N}\{y_{n+2} - \widehat{m}_{h,n}(z_n)\}^2 \\ &= \frac{1}{N}\sum_{n=1}^{N}\widehat{m}_{h,n}(z_n)^2 - \frac{2}{N}\sum_{n=1}^{N}\widehat{m}_{h,n}(z_n)y_{n+2} + \frac{1}{N}\sum_{n=1}^{N}y_{n+2}^2 \\ &\equiv CV(h)_1 + CV(h)_2 + CV(h)_3. \end{aligned} \tag{6.4.17}$$

TABLE 6.7. Simulation results for model 1 in example 6.4.1

N	$\mid \widehat{h}_C - \widehat{h}_D \mid$	$\mid \widehat{\beta}(\widehat{h}_C) - 0.25 \mid$	$\mid \widetilde{\beta}(\widehat{h}_C) - 0.25 \mid$	$ASE(\widehat{h}_C)$
100	0.08267	0.09821	0.09517	0.00451
200	0.07478	0.07229	0.07138	0.00229
300	0.08606	0.05748	0.05715	0.00108
400	0.05582	0.05526	0.05504	0.00117
500	0.07963	0.05025	0.05013	0.00076

TABLE 6.8. Simulation results for model 2 in example 6.4.1

N	$\mid \widehat{h}_C - \widehat{h}_D \mid$	$\mid \widehat{\beta}(\widehat{h}_C) - 0.25 \mid$	$\mid \widetilde{\beta}(\widehat{h}_C) - 0.25 \mid$	$ASE(\widehat{h}_C)$
100	0.08952	0.07481	0.07367	0.05246
200	0.08746	0.06215	0.06189	0.02635
300	0.09123	0.05243	0.05221	0.01573
400	0.09245	0.05138	0.05093	0.01437
500	0.09561	0.05042	0.05012	0.01108

Hence, $CV(h)_1$ and $CV(h)_2$ can be computed by the similar reason as those of D_{1h} and D_{2h}. Therefore the problem of minimizing $CV(h)$ over H_{N+2} is the same as that of minimizing $CV(h)_1 + CV(h)_2$. That is

$$\widehat{h}_C = \arg \min_{h \in H_{N+2}} \{CV(h)_1 + CV(h)_2\}. \tag{6.4.18}$$

(v). Under the cases of $T = 102, 202, 302, 402$, and 502, compute

$$|\widehat{h}_C - \widehat{h}_D|, \ |\widehat{\beta}(\widehat{h}_C) - \beta|, \ |\widetilde{\beta}(\widehat{h}_C) - \beta|, \tag{6.4.19}$$

and

$$\text{ASE}(\widehat{h}_C) = \frac{1}{N} \sum_{n=1}^{N} \{\widehat{g}^*_{\widehat{h}_C,n}(y_n) - g(y_n)\}^2. \tag{6.4.20}$$

The following simulation results were performed 1000 times using the Splus functions (Chambers and Hastie (1992)) and the means are tabulated in Tables 6.7 and 6.8.

Remark 6.4.5 *Table 6.7 gives the small sample results for the Mackey-Glass system with $(a, b, d, k) = (1/4, 1/2, 2, 2)$ (§4 of Nychka, Elliner, Gallant and McCaffrey (1992)). Table 6.8 provides the small sample results for Model 2 which contains a sin function at lag 2. Trigonometric functions have been used in the time series literature to describe periodic series. Both Tables 6.7 and 6.8 show that* *when the bandwidth parameter was proportional to the reasonable candidate* $T^{-1/5}$, *the absolute errors of the data-driven estimates* $\widehat{\beta}(\widehat{h}_C)$, $\widetilde{\beta}(\widehat{h}_C)$ *and* $ASE(\widehat{h}_C)$ *decreased as the sample size* T *increased, and* $|\widetilde{\beta}(\widehat{h}_C) - 0.25| < |\widehat{\beta}(\widehat{h}_C) - 0.25|$ *for all the sample sizes. Thus, the* CV*-based* $\widehat{h}_C$ *and the adaptive data-driven estimator* $\widetilde{\beta}(\widehat{h}_C)$ *are recommended to use in practice.*

Example 6.4.2 *In this example, we consider the Canadian lynx data. This data set is the annual record of the number of Canadian lynx trapped in the MacKenzie River district of North-West Canada for the years 1821 to 1934. Tong (1977) fitted an eleventh-order Gaussian autoregressive model to* $y_t = \log_{10}$*(number of lynx trapped in the year* $(1820+t)$*) for* $t = 1, 2, ..., 114$ *(*$T = 114$*). It follows from the definition of* $(y_t, 1 \leq t \leq 114)$ *that all the transformed values* y_t *are bounded by one.*

Several models have already been used to fit the lynx data. Tong (1977) proposed the eleventh-order Gaussian autoregressive model to fit the data. See also Tong (1990). More recently, Wong and Kohn (1996) used a second-order additive autoregressive model of the form

$$y_t = g_1(y_{t-1}) + g_2(y_{t-2}) + e_t \tag{6.4.21}$$

to fit the data, where g_1 and g_2 are smooth functions. The authors estimated both g_1 and g_2 through using a Bayesian approach and their conclusion is that the estimate of g_1 is almost linear while the estimate of g_2 is nonlinear. Their research suggests that if we choose either model (6.4.22) or model (6.4.23) below to fit the lynx data, model (6.4.22) will be more appropriate.

$$\begin{aligned} y_t &= \beta_1 y_{t-1} + g_2(y_{t-2}) + e_{1t}, && (6.4.22) \\ y_t &= g_1(y_{t-1}) + \beta_2 y_{t-2} + e_{2t}, && (6.4.23) \\ & && (6.4.24) \end{aligned}$$

where β_1 and β_2 are unknown parameters, g_1 and g_2 are unknown functions, and e_{1t} and e_{2t} are assumed to be i.i.d. random errors with zero mean and finite variance.

Our experience suggests that the choice of the kernel function is much less critical than that of the bandwidth . For Example 6.4.1, we choose the kernel function $K(x) = (2\pi)^{-1/2}\exp(-x^2/2)$, the weight function $w(x) = I_{[1,4]}(x)$, $h \in H_{114} = [0.3 \cdot 114^{-7/30}, 1.1 \cdot 114^{-1/6}]$. For model (6.4.22), similar to (6.4.6), we define the following CV function ($N = T - 2$) by

$$CV_1(h) = \frac{1}{N}\sum_{n=1}^{N}\Big(y_{n+2} - [\widetilde{\beta}_1(h)y_{n+1} + \{\widehat{g}_{1,n}(y_n) - \widetilde{\beta}_1(h)\widehat{g}_{2,n}(y_n)\}]\Big)^2,$$

where

$$\begin{aligned}\widetilde{\beta}_1(h) &= \Big\{\sum_{n=1}^{N} v_{1,n}w_{1,n}\Big\}\Big/\Big\{\sum_{n=1}^{N} v_{1,n}^2\Big\}, \quad v_{1,n} = y_{n+1} - \widehat{g}_{2,n}(y_n),\\ w_{1,n} &= y_{n+2} - \widehat{g}_{1,n}(y_n),\\ \widehat{g}_{i,n}(y_n) &= \Big\{\sum_{m=1,\neq n}^{N} K_h(y_n - y_m)y_{m+3-i}\Big\}\Big/\Big\{\sum_{m=1,\neq n}^{N} K_h(y_n - y_m)\Big\},\end{aligned}$$

for $i = 1, 2$, and $K_h(\cdot) = 1/hK(\cdot/h)$.

For model (6.4.23), we have

$$CV_2(h) = \frac{1}{N}\sum_{n=1}^{N}[y_{n+2} - \{\widehat{g}_{n,1}(y_{n+1}) - \widetilde{\beta}_2(h)\widehat{g}_{n,2}(y_{n+1}) + \widetilde{\beta}_2(h)y_n\}]^2,$$

where

$$\begin{aligned}\widetilde{\beta}_2(h) &= \Big\{\sum_{n=1}^{N} v_{2,n}w_{2,n}\Big\}\Big/\Big\{\sum_{n=1}^{N} v_{2,n}^2\Big\}, \quad v_{2,n} = y_n - \widehat{g}_{n,2}(y_{n+1}),\\ w_{2,n} &= y_{n+2} - \widehat{g}_{n,1}(y_{n+1}),\\ \widehat{g}_{n,i}(y_{n+1}) &= \Big\{\sum_{m=1,\neq n}^{N} K_h(y_{n+1} - y_{m+1})y_{m+2i(2-i)}\Big\}\Big/\Big\{\sum_{m=1,\neq n}^{N} K_h(y_{n+1} - y_{m+1})\Big\}\end{aligned}$$

for $i = 1, 2$.

Through minimizing the CV functions $CV_1(h)$ and $CV_2(h)$, we obtain

$$CV_1(\widehat{h}_{1C}) = \inf_{h \in H_{114}} CV_1(h) = 0.0468 \text{ and } CV_2(\widehat{h}_{2C}) = \inf_{h \in H_{114}} CV_2(h) = 0.0559$$

respectively. The estimates of the error variance of $\{e_{1t}\}$ and $\{e_{2t}\}$ were 0.04119 and 0.04643 respectively. The estimate of the error variance of the model of Tong (1977) was 0.0437, while the estimate of the error variance of the model of Wong and Kohn (1996) was 0.0421 which is comparable with our variance estimate of 0.04119. Obviously, the approach of Wong and Kohn cannot provide explicit estimates for f_1 and f_2 since their approach depends heavily on the Gibbs sampler. Our CPU time for Example 6.4.2 took about 30 minutes on a Digital workstation. Time plot of the common-log-transformed lynx data (part (a)), full plot of fitted values (solid) and the observations (dashed) for model (6.4.22) (part (b)), partial plot of the LS estimator ($1.354y_{n+1}$) against y_{n+1} (part (c)), partial plot of the nonparametric estimate ($\widetilde{g}_2$) of g_2 in model (6.4.22) against y_n (part (d)), partial plot of the nonparametric estimate of g_1 in model (6.4.23) against y_{n+1} (part (e)), and partial plot of the LS estimator ($-0.591y_n$) against y_n (part (f)) are given in Figure 6.1 on page 180. For the Canadian lynx data, when selecting y_{t-1} and y_{t-2} as the candidates of the regressors, our research suggests using the following prediction equation

$$\widehat{y}_{n+2} = 1.354y_{n+1} + \widetilde{g}_2(y_n), \ n = 1, 2, \ldots, \tag{6.4.25}$$

where

$$\widetilde{g}_2(y_n) = \widehat{g}_1(y_n, \widehat{h}_{1C}) - 1.354\widehat{g}_2(y_n, \widehat{h}_{1C})$$

and

$$\widehat{g}_i(y_n, h) = \Big\{\sum_{m=1}^{N} K_h(y_n - y_m)y_{m+3-i}\Big\} \Big/ \Big\{\sum_{m=1}^{N} K_h(y_n - y_m)\Big\},$$

in which $i = 1, 2$ and $\widehat{h}_{1C} = 0.1266$. Part (d) of Figure 6.1 shows that $\widetilde{g}_2$ appears to be nonlinear.

We now compare the methods of Tong (1977), Wong and Kohn (1996) and our approach. The research of Wong and Kohn (1996) suggests that for the lynx data, the second-order additive autoregressive model (6.4.25) is more reasonable than the threshold autoregressive method proposed by Tong (1977). It follows from (6.4.25) that for the lynx data the partially linear autoregressive model of the form (6.4.1) is easier to implement in practice than the second-order additive autoregressive model of Wong and Kohn (1996).

Remark 6.4.6 *This section mainly establishes the estimation procedure for model (6.4.1). As mentioned in Remarks 6.4.2 and 6.4.3, however, the estimation procedure can be extended to cover a broad class of models. Section 6.4.2 demonstrates that the estimation procedure can be applied to both simulated and real data examples. A software for the estimation procedure is available upon request. This section shows that semiparametric methods can not only retain the beauty of linear regression methods but provide 'models' with better predictive power than is available from nonparametric methods.*

6.5 Other Related Developments

In Sections 6.2–6.4, we have discussed the parametric and nonparametric tests, the optimum linear subset selection and the optimal bandwidth parameter selection for model (6.4.1). In this section, we summarize recent developments in a general class of additive stochastic regression models including model (6.2.1).

Consider the following additive stochastic regression model

$$Y_t = m(X_t) + e_t = \sum_{i=1}^{p} g_i(U_{ti}) + g(V_t) + e_t, \qquad (6.5.1)$$

where $X_t = (U_t^T, V_t^T)^T$, $U_t = (U_{t1}, \ldots, U_{tp})^T$, $V_t = (V_{t1}, \ldots, V_{td})^T$, and g_i are unknown functions on R^1. For $Y_t = y_{t+r}$, $U_{ti} = V_{ti} = y_{t+r-i}$ and $g_i(U_{ti}) = \beta_i U_{ti}$, model (6.5.1) is a semiparametric AR model discussed in Section 6.2. For $Y_t = y_{t+r}$, $U_{ti} = y_{t+r-i}$ and $g \equiv 0$, model (6.5.1) is an additive autoregressive model discussed extensively by Chen and Tsay (1993). Recently, Masry and Tjøstheim (1995, 1997) discussed nonlinear ARCH time series and an additive nonlinear bivariate ARX model, and proposed several consistent estimators. See Tjøstheim (1994), Tong (1995) and Härdle, Lütkepohl and Chen (1997) for recent developments in nonlinear and nonparametric time series models. Recently, Gao, Tong and Wolff (1998a) considered the case where $g \equiv 0$ in (6.5.1) and discussed the lag selection and order determination problem. See Tjøstheim and Auestad (1994a, 1994b) for the nonparametric autoregression case and Cheng and Tong (1992, 1993) for the stochastic dynamical systems case. More recently, Gao, Tong and Wolff (1998b) proposed an adaptive test statistic for testing additivity for the case where each g_i in (6.5.1) is an unknown function in R^1. Asymptotic theory and power investigations of the test statistic have been discussed under some mild conditions. This research generalizes the discussion of Chen, Liu and Tsay (1995) and Hjellvik and Tjøstheim (1995) for testing additivity and linearity in nonlinear autoregressive models. See also Kreiss, Neumann and Yao (1997) for Bootstrap tests in nonparametric time series regression.

Further investigation of (6.5.1) is beyond the scope of this monograph. Recent developments can be found in Gao, Tong and Wolff (1998a, b).

6.6 The Assumptions and the Proofs of Theorems

6.6.1 Mathematical Assumptions

Assumption 6.6.1 *(i) Assume that $\{e_t\}$ is a sequence of i.i.d. random processes with $Ee_t = 0$ and $Ee_t^2 = \sigma_0^2 < \infty$, and that e_s are independent of X_t for all $s \geq t$.*

(ii) Assume that X_t are strictly stationary and satisfy the Rosenblatt mixing condition

$$\sup\{|P(A \cap B) - P(A)P(B)| : A \in \Omega_1^l, B \in \Omega_{l+k}^{\infty}\} \leq C_1 \exp(-C_2 k)$$

for all $l, k \geq 1$ and for constants $\{C_i > 0 : i = 1, 2\}$, where $\{\Omega_i^j\}$ denotes the σ-field generated by $\{X_t : i \leq t \leq j\}$.

Let $g^{(m)}$ be the m-order derivative of the function g and M be a constant,

$$G_m(S) = \{g : |g^{(m)}(s) - g^{(m)}(s')| \leq M||s - s'||\},$$

where m is an integer, $s, s' \in S$, a compact subset of R^d, $0 < M < \infty$, and $||\cdot||$ denotes the Euclidean norm.

Assumption 6.6.2 *For $g \in G_m(S)$ and $\{z_j(\cdot) : j = 1, 2, \ldots\}$ given above, there exists a vector of unknown parameters $\gamma = (\gamma_1, \ldots, \gamma_q)^T$ such that for a constant C_0 $(0 < C_0 < \infty)$ independent of T*

$$q^{2(m+1)} E\Big\{\sum_{j=1}^{q} z_j(V_t)\gamma_j - g(V_t)\Big\}^2 \sim C_0$$

where the symbol "$\sim$" indicates that the ratio of the left-hand side and the right-hand side tends to one as $T \to \infty$, $q = [l_0 T^{\frac{1}{2(m+1)+1}}]$, in which $0 < l_0 < \infty$ is a constant.

Assumption 6.6.3 *(i) Z is of full column rank q, $\{z_i(\cdot) : 1 \leq i \leq q\}$ is a sequence of continuous functions with $\sup_v \sup_{i \geq 1} |z_i(v)| < \infty$.*

(ii) Assume that $d_i^2 = E\{z_i(V_t)^2\}$ exist with the absolute constants d_i^2 satisfying $0 < d_1^2 \leq \cdots \leq d_q^2 < \infty$ and that

$$E\{z_i(V_t)z_j(V_t)\} = 0 \text{ and } E\{z_k(V_s)z_k(V_t)\} = 0$$

for all $i \neq j$, $k \geq 1$ and $s \neq t$.

(iii) Assume that $c_i^2 = E\{U_{ti}^2\}$ *exist with the constants* c_i^2 *satisfying* $0 < c_1^2 \leq \cdots \leq c_p^2 < \infty$ *and that the random processes* $\{U_t : t \geq 1\}$ *and* $\{z_k(V_t) : k \geq 1\}$ *satisfy the following orthogonality conditions*

$$E\{U_{ti}U_{tj}\} = 0 \text{ and } E\{U_{si}U_{tj}z_k(V_s)z_k(V_t)\} = 0$$

for all $i \neq j$, $k \geq 1$ *and* $s \neq t$.

Assumption 6.6.4 *There exists an absolute constant* $M_0 \geq 4$ *such that for all* $t \geq 1$

$$\sup_x E\{|Y_t - E(Y_t|X_t)|^{2M_0}|X_t = x\} < \infty.$$

Assumption 6.6.5 *(i)* K_d *is a* d*-dimensional symmetric, Lipschitz continuous probability kernel function with* $\int ||u||^2 K_d(u)du < \infty$, *and has an absolutely integrable Fourier transform, where* $||\cdot||$ *denotes the Euclidean norm.*

(ii) The distribution of X_t *is absolutely continuous, and its density* f_X *is bounded below by* c_f *and above by* d_f *on the compact support of* f_X.

(iii) The density function $f_{V,A}$ *of random vector* V_{tA} *has a compact support on which all the second derivatives of* $f_{V,A}$, g_{1A} *and* g_{2A} *are continuous, where* $g_{1A}(v) = E(Y_t|V_{tA} = v)$ *and* $g_{2A}(v) = E(U_{tA}|V_{tA} = v)$.

Assumption 6.6.6 *The true regression function* $U_{tA_0}^T\beta_{A_0} + g_{A_0}(V_{tA_0})$ *is unknown and nonlinear.*

Assumption 6.6.7 *Assume that the lower and the upper bands of* H_T *satisfy*

$$\lim_{T\to\infty} h_{\min}(T,d)T^{1/(4+d_0)+c_1} = a_1 \text{ and } \lim_{T\to\infty} h_{\max}(T,d)T^{1/(4+d_0)-c_1} = b_1,$$

where $d_0 = r - |A_0|$, *the constants* a_1, b_1 *and* c_1 *only depend on* (d, d_0) *and satisfy* $0 < a_1 < b_1 < \infty$ *and* $0 < c_1 < 1/\{4(4+d_0)\}$.

Assumption 6.6.8 *(i) Assume that* y_t *are strictly stationary and satisfy the Rosenblatt mixing condition.*

(ii) K *is symmetric, Lipschitz continuous and has an absolutely integrable Fourier transform.*

(iii) K *is a bounded probability kernel function with* $\int_{-\infty}^{\infty} u^2 K(u)du < \infty$.

(iv) Assume that the weight function w is bounded and that its support S is compact.

(v) Assume that $\{y_t\}$ has a common marginal density $f(\cdot)$, $f(\cdot)$ has a compact support containing S, and $g_i(\cdot)$ $(i = 1, 2)$ and $f(\cdot)$ have two continuous derivatives on the interior of S.

(vi) For any integer $k \geq 1$, $E|y_t|^k < \infty$.

Remark 6.6.1 *(i) Assumption 6.6.1(i) can be replaced by a more complicated condition that includes the conditional heteroscedasticity case. Details can be found in Gao, Tong and Wolff (1998b).*

(ii) Assumption 6.6.1(ii) is quite common in such problems. See, for example, (C.8) in Härdle and Vieu (1992). However, it would be possible, but with more tedious proofs, to obtain the above Theorems under less restrictive assumptions that include some algebraically decaying rates.

(iii) As mentioned before, Assumptions 6.6.2 and 6.6.3 provide some smooth and orthogonality conditions. In almost all cases, they hold if $g(\cdot)$ satisfies some smoothness conditions. In particular, they hold when Assumption 6.6.1 holds and the series functions are either the family of trigonometric series or the Gallant's (1981) flexible Fourier form. Recent developments in nonparametric series regression for the i.i.d. case are given in Andrews (1991) and Hong and White (1995)

Remark 6.6.2 *(i) Assumption 6.6.5(ii) guarantees that the model we consider is identifiable, which implies that the unknown parameter vector β_{A_0} and the true nonparametric component $g_{A_0}(V_{tA_0})$ are uniquely determined up to a set of measure zero.*

(ii) Assumption 6.6.6 is imposed to exclude the case where $g_{A_0}(\cdot)$ is also a linear function of a subset of $X_t = (X_{t1}, \ldots, X_{tr})^T$.

(iii) Assumption 6.6.8 is similar to Assumptions 6.6.1 and 6.6.5. For the sake of convenience, we list all the necessary conditions for Section 6.4 in the separate assumption–Assumption 6.6.8.

6.6.2 Technical Details

Proof of Theorem 6.2.1

For simplicity, let C_i $(0 < |C_i| < \infty)$ denote positive constants which may have different values at each appearance throughout this section. Before proving Theorem 6.2.1, we state a lemma, whose proof can be found in Lemma A.1 of Gao, Tong and Wolff (1998a).

Lemma 6.6.1 *Assume that the conditions of Theorem 6.2.1 hold. Then*

$$c_1^2 + o_P(\delta_1(p)) \le \lambda_{\min}(\frac{1}{T}U^TU) \le \lambda_{\max}(\frac{1}{T}U^TU) \le c_p^2 + o_P(\delta_1(p))$$
$$d_1^2 + o_P(\lambda_2(q)) \le \lambda_{\min}(\frac{1}{T}Z^TZ) \le \lambda_{\max}(\frac{1}{T}Z^TZ) \le d_q^2 + o_P(\lambda_2(q))$$

where $c_i^2 = EU_{ti}^2$, *and for all* $i = 1, 2, \ldots, p$ *and* $j = 1, 2, \ldots, q$

$$\lambda_i\Big\{\frac{1}{T}U^TU - I_1(p)\Big\} = o_P(\delta_1(p)),$$
$$\lambda_j\Big\{\frac{1}{T}Z^TZ - I_2(q)\Big\} = o_P(\lambda_2(q)),$$

where

$$I_1(p) = diag(c_1^2, \ldots, c_p^2) \text{ and } I_2(q) = diag(d_1^2, \ldots, d_q^2)$$

are $p\times p$ *and* $q\times q$ *diagonal matrices respectively,* $\lambda_{\min}(B)$ *and* $\lambda_{\max}(B)$ *denote the smallest and largest eigenvalues of matrix* B, $\{\lambda_i(D)\}$ *denotes the* i*–th eigenvalue of matrix* D, *and* $\delta_1(p) > 0$ *and* $\lambda_2(q) > 0$ *satisfy as* $T \to \infty$, $\max\{\delta_1(p), \lambda_2(q)\} \cdot \max\{p, q\} \to 0$ *as* $T \to \infty$.

The Proof of Theorem 6.2.1. Here we prove only Theorem 6.2.1(i) and the second part follows similarly. Without loss of generality, we assume that the inverse matrices $(Z^TZ)^{-1}$, $(U^TU)^{-1}$ and $(\widehat{U}^T\widehat{U})^{-1}$ exist and that $d_1^2 = d_2^2 = \cdots = d_q^2 = 1$ and $\sigma_0^2 = 1$.

By (6.2.7) and Assumption 6.6.2, we have

$$\begin{aligned}(Z^TZ)^{1/2}(\widehat{\gamma} \; - \; \gamma) &= (Z^TZ)^{-1/2}Z^T\{\mathcal{F} - U(\widehat{U}^T\widehat{U})^{-1}\widehat{U}^T\}(e+\delta)\\ &= (Z^TZ)^{-1/2}Z^Te + (Z^TZ)^{-1/2}Z^T\delta - (Z^TZ)^{-1/2}Z^TU(\widehat{U}^T\widehat{U})^{-1}\widehat{U}^Te\\ &\quad -(Z^TZ)^{-1/2}Z^TU(\widehat{U}^T\widehat{U})^{-1}\widehat{U}^T\delta\\ &\equiv I_{1T} + I_{2T} + I_{3T} + I_{4T}, \end{aligned} \tag{6.6.1}$$

where $e = (e_1, \ldots, e_T)^T$, $\delta = (\delta_1, \ldots, \delta_T)^T$, and $\delta_t = g(V_t) - Z(V_t)^T\gamma$.

In view of (6.6.1), in order to prove Theorem 6.2.1, it suffices to show that

$$(2q)^{-1/2}(e^T Pe - q) \longrightarrow^{\mathcal{L}} N(0,1) \tag{6.6.2}$$

and for $i = 2, 3, 4$

$$I_{1T}^T I_{iT} = o_P(q^{1/2}) \text{ and } I_{iT}^T I_{iT} = o_P(q^{1/2}). \tag{6.6.3}$$

Before proving (6.6.2), we need to prove

$$e^T Pe = \sum_{1\leq s,t\leq T} a_{st}e_s e_t + o_P(q^{1/2}), \tag{6.6.4}$$

where $a_{st} = 1/T \sum_{i=1}^q z_i(V_s)z_i(V_t)$.

In order to prove (6.6.4), it suffices to show that

$$q^{-1/2}|e^T(P - P_0)e| = o_P(1), \tag{6.6.5}$$

where $P_0 = \{a_{st}\}_{1\leq s,t\leq T}$ is a matrix of order $T \times T$.

Noting that Lemma 6.6.1 holds and

$$\begin{aligned}
\Big|e^T Z(Z^T Z)^{-1}\{I_2(q) &- \frac{1}{T}Z^T Z\}I_2(q)^{-1}Z^T e\Big|^2 \\
&\leq e^T Z(Z^T Z)^{-1}Z^T e \\
&\times e^T Z\{I_2(q) - \frac{1}{T}Z^T Z\}(Z^T Z)^{-1}\{I_2(q) - \frac{1}{T}Z^T Z\}Z^T e \\
&\leq \frac{C}{T^2}\lambda_{\max}\{(I_2(q) - \frac{1}{T}Z^T Z)^2\}(e^T Z Z^T e)^2,
\end{aligned}$$

in order to prove (6.6.5), it suffices to show that

$$\lambda_2(q)T^{-1}q^{-1/2}(Z^T e)^T(Z^T e) = o_P(1), \tag{6.6.6}$$

which follows from Markov inequality and

$$\begin{aligned}
P\{\lambda_2(q)T^{-1}q^{-1/2}(Z^T e)^T(Z^T e) > \varepsilon\} &\leq \varepsilon^{-1}\lambda_2(q)T^{-1}q^{-1/2}E\sum_{i=1}^q\Big\{\sum_{t=1}^T z_i(V_t)e_t\Big\}^2 \\
&\leq C\lambda_2(q)T^{-1}q^{-1/2}qT \\
&= C\lambda_2(q)q^{1/2} = o(1)
\end{aligned}$$

using Assumptions 6.6.1 and 6.6.3. Thus, the proof of (6.6.4) is completed.

Noting (6.6.4), in order to prove (6.6.2), it suffices to show that as $T \to \infty$

$$q^{-1/2}\Big(\sum_{t=1}^{T} a_{tt}e_t^2 - q\Big) \longrightarrow_P 0 \tag{6.6.7}$$

and

$$\sum_{t=2}^{T} W_{Tt} \longrightarrow^{\mathcal{L}} N(0,1), \tag{6.6.8}$$

where $W_{Tt} = (2/q)^{1/2} \sum_{s=1}^{t-1} a_{st}e_s e_t$ forms a zero mean martingale difference.

Now applying a central limit theorem for martingale sequences (see Theorem 1 of Chapter VIII of Pollard (1984)), we can deduce

$$\sum_{t=2}^{T} W_{Tt} \longrightarrow^{\mathcal{L}} N(0,1)$$

if

$$\sum_{t=2}^{T} E(W_{Tt}^2|\Omega_{t-1}) \longrightarrow_{\mathcal{P}} 1 \tag{6.6.9}$$

and

$$\sum_{t=2}^{T} E\{W_{Tt}^2 I(|W_{Tt}| > c)|\Omega_{t-1}\} \longrightarrow_{\mathcal{P}} 0 \tag{6.6.10}$$

for all $c > 0$.

It is obvious that in order to prove (6.6.9) and (6.6.10), it suffices to show that as $T \to \infty$

$$\frac{2}{q}\sum_{t=2}^{T}\sum_{s=1}^{t-1} a_{st}^2 e_s^2 - 1 \longrightarrow_{\mathcal{P}} 0, \tag{6.6.11}$$

$$\frac{2}{q}\sum_{t=1}^{T}\sum_{r\neq s} a_{st}a_{rt}e_s e_r \longrightarrow_{\mathcal{P}} 0, \tag{6.6.12}$$

and

$$\frac{4}{q^2}\sum_{t=2}^{T} E\left(\sum_{s=1}^{t-1} a_{st}e_s\right)^4 \to 0. \tag{6.6.13}$$

The left-hand side of (6.6.11) is

$$\frac{2}{q}\sum_{t=2}^{T}\sum_{s=1}^{t-1} a_{st}^2(e_s^2 - 1) + \left(\frac{2}{q}\sum_{t=2}^{T}\sum_{s=1}^{t-1} a_{st}^2 - 1\right). \tag{6.6.14}$$

Also, the first term in (6.6.14) is

$$\frac{2}{q}\sum_{i=1}^{q}\frac{1}{T}\Big\{\sum_{t=2}^{T}z_i(V_t)^2\times\frac{1}{T}\sum_{s=1}^{t-1}z_i(V_s)^2(e_s^2-1)\Big\}$$
$$+\frac{2}{q}\sum_{i\neq j=1}^{q}\Big\{\frac{1}{T^2}\sum_{t=2}^{T}\sum_{s=1}^{t-1}f_{ij}(V_s,V_t)(e_s^2-1)\Big\}$$
$$\equiv\frac{2}{q}\sum_{i=1}^{q}\Big\{\frac{1}{T}\sum_{t=2}^{T}z_i(V_t)^2\cdot M_{1t}(i)\Big\}+\frac{2}{q}\sum_{i=1}^{q}\sum_{j=1,\neq i}^{q}M_{1ij}, \tag{6.6.15}$$

where $f_{ij}(V_s,V_t)=z_i(V_s)z_j(V_s)z_i(V_t)z_j(V_t)$.

The second term in (6.6.14) is

$$\frac{1}{q}\sum_{i=1}^{q}\Big[\Big\{\frac{1}{T}\sum_{t=1}^{T}z_i(V_t)^2-1\Big\}^2+2\Big\{\frac{1}{T}\sum_{t=1}^{T}z_i(V_t)^2-1\Big\}\Big]$$
$$+\frac{1}{q}\sum_{i=1}^{q}\sum_{j=1,\neq i}^{q}\frac{1}{T^2}\sum_{t=1}^{T}\sum_{s=1}^{T}z_i(V_s)z_j(V_s)z_i(V_t)z_j(V_t)$$
$$\equiv\frac{1}{q}\sum_{i=1}^{q}M_i+\frac{1}{q}\sum_{i=1}^{q}\sum_{j=1,\neq i}^{q}M_{2ij}. \tag{6.6.16}$$

Analogously, the left-hand side of (6.6.12) is

$$\frac{4}{q}\sum_{i=1}^{q}\frac{1}{T}\sum_{t=2}^{T}z_i(V_t)^2\times\frac{1}{T}\sum_{r=2}^{t-1}\sum_{s=1}^{r-1}z_i(V_s)z_i(V_r)e_se_r$$
$$+\frac{4}{q}\sum_{i=1}^{q}\sum_{j=1,\neq i}^{q}\frac{1}{T^2}\sum_{t=2}^{T}\sum_{r=2}^{t-1}\sum_{s=1}^{r-1}g_{ij}(V_s,V_r,V_t)$$
$$\equiv\frac{4}{q}\sum_{i=1}^{q}\frac{1}{T}\sum_{t=2}^{T}z_i(V_t)^2\cdot M_{2t}(i)+\frac{4}{q}\sum_{i=1}^{q}\sum_{j=1,\neq i}^{q}M_{3ij}, \tag{6.6.17}$$

where $g_{ij}(V_s,V_r,V_t)=z_i(V_s)e_sz_j(V_r)e_rz_i(V_t)z_j(V_t)$.

Using Assumptions 6.6.1-6.6.3 and applying the martingale limit results of Chapters 1 and 2 of Hall and Heyde (1980), we can deduce for $i=1,2$, $j=1,2,3$ and $s,r\geq 1$

$$M_s=o_P(1),\ M_{jsr}=o_P(q^{-1}),$$

and

$$\max_{1\leq t\leq T}|M_{it}(s)|=o_P(1). \tag{6.6.18}$$

Thus, equations (6.6.11) and (6.6.12) follows from (6.6.14)–(6.6.18). Now, we begin to prove (6.6.13). Obviously,

$$\Big(\sum_{s=1}^{t-1}a_{st}e_s\Big)^4=\frac{1}{T^4}\Big\{\sum_{i=1}^{q}\sum_{s=1}^{t-1}z_i(V_s)e_sz_i(V_t)\Big\}^4\leq\frac{Cq^3}{T^4}\sum_{i=1}^{q}\Big\{\sum_{s=1}^{t-1}z_i(V_s)e_s\Big\}^4 \tag{6.6.19}$$

using Assumption 6.6.3(i).

For any fixed $i \geq 1$

$$\begin{aligned} \sum_{t=2}^{T} E\Big\{\sum_{s=1}^{t-1} z_i(V_s)e_s\Big\}^4 &= C_1 \sum_{t=2}^{T}\sum_{s=1}^{t-1} E\{z_i(V_s)e_s\}^4 \\ &\quad + C_2 \sum_{t=2}^{T}\sum_{s_1=1}^{t-1}\sum_{s_2=1,\neq s_1}^{t-1} E\{z_i(V_{s_1})^2 e_{s_1}^2 z_i(V_{s_2})^2 e_{s_2}^2\} \\ &\equiv J_{i1T} + J_{i2T}. \end{aligned}$$

Applying Assumptions 6.6.1(i) and 6.6.3(iii) again, we have for $j = 1, 2$

$$J_{ijT} \leq C_3 T^3. \tag{6.6.20}$$

Thus, (6.6.19)–(6.6.20) imply (6.6.13). We have therefore proved equations (6.6.11) –(6.6.13). As a result, equation (6.6.8) holds. In the following, we begin to prove (6.6.7).

For any $1 \leq i \leq q$, define

$$\phi_{ii} = \frac{1}{T}\sum_{t=1}^{T}\{z_i(V_t)^2 - 1\}.$$

Using Assumptions 6.6.1 and 6.6.3, and applying Lemma 3.2 of Boente and Fraiman (1988), we have for any given $\varepsilon > 0$

$$\begin{aligned} P(\max_{1\leq i\leq q}|\phi_{ii}| > \varepsilon q^{-1/2}) &\leq \sum_{i=1}^{q} P(|\phi_{ii}| > \varepsilon q^{-1/2}) \\ &\leq C_1 q \exp(-C_2 T^{1/2} q^{-1/2}) \to 0 \end{aligned}$$

as $T \to \infty$, where C_i are constants.

Thus

$$\max_{1\leq i\leq q}|\phi_{ii}| = o_P(q^{-1/2}).$$

Therefore, using Assumptions 6.6.1 and 6.6.3, and applying Cauchy-Schwarz inequality, we obtain

$$\begin{aligned} \sum_{t=1}^{T} a_{tt}e_t^2 - q &= \sum_{t=1}^{T} a_{tt}(e_t^2 - 1) + tr[Z\{1/T I_2(q)^{-1}\}Z^T - P] \\ &= \sum_{t=1}^{T} a_{tt}(e_t^2 - 1) + tr[I_2(q)^{-1}\{1/T Z^T Z - I_2(q)\}] \\ &= \sum_{t=1}^{T} a_{tt}(e_t^2 - 1) + \sum_{i=1}^{q}\phi_{ii} = o_P(q^{1/2}), \end{aligned}$$

where $\mathrm{tr}(B)$ denotes the trace of matrix B.

The proof of (6.6.2) is consequently completed. Before proving (6.6.3), we need to prove for T large enough

$$\lambda_{\max}(\frac{1}{T}ZZ^T) = o_P(q^{-1/2}),$$

which follows from

$$\begin{aligned} P\Big\{\Big|\sum_{t=1}^T\sum_{s=1}^T\sum_{i=1}^q z_i(V_s)z_i(V_t)l_sl_t\Big| &> \varepsilon Tq^{-1/2}\Big\} \\ &\le \frac{q^{1/2}}{T\varepsilon}E\Big|\sum_{t=1}^T\sum_{s=1}^T\sum_{i=1}^q z_i(V_s)z_i(V_t)l_sl_t\Big| \\ &\le C\frac{q^{1/2}}{T}\sum_{i=1}^q E\Big|\sum_{t=1}^T\sum_{s=1}^T z_i(V_s)z_i(V_t)l_sl_t\Big| \\ &= C\frac{q^{1/2}}{T}\sum_{i=1}^q E\Big\{\sum_{t=1}^T z_i(V_t)l_t\Big\}^2 \le C\frac{q^{3/2}}{T}\to 0 \end{aligned}$$

using Assumption 6.6.3(ii) and the fact that $l=(l_1,\ldots,l_T)^T$ is any identical vector satisfying $\sum_{t=1}^T l_t^2=1$. Thus

$$\delta^TP\delta \le \lambda_{\max}\{(Z^TZ)^{-1}\}\delta^TZZ^T\delta \le \frac{C}{T}\lambda_{\max}(ZZ^T)\delta^T\delta = o_P(q^{1/2}).$$

We now begin to prove

$$I_{3T}^TI_{3T} = o_P(q^{1/2}).$$

It is obvious that

$$\begin{aligned} I_{3T}^TI_{3T} &= e^T\widehat{U}(\widehat{U}^T\widehat{U})^{-1}U^TPU(\widehat{U}^T\widehat{U})^{-1}\widehat{U}^Te \\ &\le \lambda_{\max}(U^TPU)\cdot\lambda_{\max}\{(\widehat{U}^T\widehat{U})^{-1}\}\cdot e^T\widehat{U}(\widehat{U}^T\widehat{U})^{-1}\widehat{U}^Te \\ &\le \frac{C}{T}\lambda_{\max}\{(Z^TZ)^{-1}\}\lambda_{\max}(U^TZZ^TU)e^T\widehat{U}(\widehat{U}^T\widehat{U})^{-1}\widehat{U}^Te. \quad (6.6.21) \end{aligned}$$

Similar to the proof of (6.6.4), we can prove

$$e^T\widehat{U}(\widehat{U}^T\widehat{U})^{-1}\widehat{U}^Te = C_1p + o_P(p^{1/2}) \quad (6.6.22)$$

using Assumptions 6.6.2 and 6.6.3.

In order to estimate the order of (6.6.21), it suffices to estimate

$$\lambda_{\max}\Big\{\frac{1}{T}(Z^TU)^T(Z^TU)\Big\}.$$

Analogous to the proof of Lemma 6.6.1, we have for $1 \leq i \leq p$

$$\lambda_i\Big\{\frac{1}{T}(Z^TU)^T(Z^TU)\Big\} \leq 2p \max_{1\leq i\neq j\leq p} |d_{ij}| = o_P(\varepsilon(p)Tq), \tag{6.6.23}$$

where

$$d_{ij} = \frac{1}{T}\sum_{l=1}^{q}\Big\{\sum_{s=1}^{T} U_{si}z_l(V_s)\Big\}\Big\{\sum_{t=1}^{T} U_{tj}z_l(V_t)\Big\}$$

denote the (i, j) elements of the matrix $1/T(Z^TU)^T(Z^TU)$ and $\varepsilon(p)$ satisfies

$$\varepsilon(p) \to 0$$

as $T \to \infty$.

Therefore, equations (6.6.21), (6.6.22) and (6.6.23) imply

$$I_{3T}^T I_{3T} \leq o_P(\varepsilon(p)qp) = o_P(q^{1/2})$$

when $\varepsilon(p) = (p\sqrt{q})^{-1}$.

Analogously, we can prove the rest of (6.6.2) similarly and therefore we finish the proof of Theorem 6.2.1(i).

Proofs of Theorems 6.3.1 and 6.3.2

Technical lemmas

Lemma 6.6.2 *Assume that Assupmtions 6.3.1, 6.6.1, and 6.6.4-6.6.6 hold,*

$$\lim_{T\to\infty} \max_{d} h_{\min}(T, d) = 0 \text{ and } \lim_{T\to\infty} \min_{d} h_{\max}(T, d)T^{1/(4+d)} = \infty.$$

Then

$$\lim_{T\to\infty} \Pr(\widehat{A}_0 \subset A_0) = 1.$$

Lemma 6.6.3 *Under the conditions of Theorem 6.3.1, we have for every given A*

$$CV(A) = \inf_{h\in H_T} CV(h, A) = \widehat{\sigma}_T^2 + C_1(A)T^{-4/(4+d)} + o_P(T^{-4/(4+d)}), \tag{6.6.24}$$

where $\widehat{\sigma}_T^2 = 1/T\sum_{t=1}^T e_t^2$ and $C_1(A)$ is a positive constant depending on A.

The following lemmas are needed to complete the proof of Lemmas 6.6.2 and 6.6.3.

Lemma 6.6.4 *Under the conditions of Lemma 6.6.2, we have for every given A and any given compact subset G of R^d*

$$\sup_{h\in H_T}\sup_{v\in G}||\widehat{g}_1(v;h,A)-g_{1A_0}(v)||=o_P(T^{-1/4}),$$
$$\sup_{h\in H_T}\sup_{v\in G}|\widehat{g}_2(v;h,A)-g_{2A_0}(v)|=o_P(T^{-1/4}),$$
$$\sup_{h\in H_T}\sup_{v\in G}|\widehat{f}(v;h,A)-f_{A_0}(v)|=o_P(T^{-1/4}).$$

where $\widehat{g}_i$ and $\widehat{f}$ are defined in (6.3.3) and (6.3.4).

Proof. The proof of Lemma 6.6.4 follows similarly from that of Lemma 1 of Härdle and Vieu (1992).

Lemma 6.6.5 *Under the conditions of Lemma 6.6.2, for every given A, the following holds uniformly over $h \in H_T$*

$$\widehat{\beta}(h,A)-\beta_{A_0}=O_P(T^{-1/2}). \tag{6.6.25}$$

Proof. The proof of (6.6.25) follows directly from the definition of $\widehat{\beta}(h,A)$ and the conditions of Lemma 6.6.5.

Proofs of Lemmas 6.6.2 and 6.6.3

Let

$$\begin{aligned}
\bar{D}_1(h,A) &= \frac{1}{T}\sum_{t=1}^{T}\{\widehat{g}_{1t}(V_{tA},h)-g_{1A_0}(V_{tA_0})\}^2,\\
\bar{D}_2(h,A) &= \frac{1}{T}\sum_{t=1}^{T}\{\widehat{g}_{2t}(V_{tA},h)-g_{2A_0}(V_{tA_0})\}\{\widehat{g}_{2t}(V_{tA},h)-g_{2A_0}(V_{tA_0})\}^T,
\end{aligned}$$

where $g_{1A_0}(V_{tA_0})=E(Y_t|V_{tA_0})$ and $g_{2A_0}(V_{tA_0})=E(U_{tA_0}|V_{tA_0})$.

Obviously,

$$\begin{aligned}
D(h,A) &= \frac{1}{T}\sum_{t=1}^{T}\{U_{tA}^T\widehat{\beta}(h,A)-U_{tA_0}^T\beta_{A_0}\}^2+\frac{1}{T}\sum_{t=1}^{T}\{\widehat{g}_t(V_{tA},\widehat{\beta}(h,A))-g_{A_0}(V_{tA_0})\}^2\\
&\quad -\frac{2}{T}\sum_{t=1}^{T}\{U_{tA}^T\widehat{\beta}(h,A)-U_{tA_0}^T\beta_{A_0}\}\{\widehat{g}_t(V_{tA},\widehat{\beta}(h,A))-g_{A_0}(V_{tA_0})\}\\
&\equiv D_1(h,A)+D_2(h,A)+D_3(h,A).
\end{aligned}$$

Similarly, we can prove

$$D_2(h,A)=\bar{D}_1(h,A)+\beta_{A_0}^T\bar{D}_2(h,A)\beta_{A_0}+o_P(D_2(h,A)).$$

and for $i = 1, 3$ and every given $A \subset \mathcal{A}$

$$\sup_{h \in H_T} \frac{D_i(h, A)}{D_2(h, A)} = o_P(1).$$

From the definition of $CV(h, A)$, we have

$$\begin{aligned} CV(h, A) &= \frac{1}{T}\sum_{t=1}^{T}\{Y_t - U_{tA}^T\widehat{\beta}(h, A) - \widehat{g}_t(V_{tA}, \widehat{\beta}(h, A))\}^2 \\ &= \frac{1}{T}\sum_{t=1}^{T} e_t^2 + D(h, A) + R(h, A), \end{aligned} \quad (6.6.26)$$

where

$$R(h, A) = \frac{2}{T}\sum_{t=1}^{T}[\{U_{tA_0}^T\beta_{A_0} + g_{A_0}(V_{tA_0})\} - \{U_{tA}^T\beta_A + \widehat{g}_t(V_{tA}, \widehat{\beta}(h, A))\}]e_t$$

satisfies for every given $A \subset \mathcal{A}$

$$\sup_{h \in H_T} \frac{R(h, A)}{D(h, A)} = o_P(1).$$

Thus, we have for every given $h \in H_T$ and $A \subset \mathcal{A}$

$$CV(h, A) = \frac{1}{T}\sum_{t=1}^{T} e_t^2 + D(h, A) + o_P(D(h, A)).$$

In order to prove Lemma 6.6.2, it suffices to show that there exists a $h^* \in H_T$ such that for any $h \in H_T$, $A \neq A_0$ and $A \subset \mathcal{A} - A_0$

$$CV(h^*, A_0) < CV(h, A),$$

which can be proved by comparing the corresponding terms of $CV(h, A_0)$ and $CV(h, A)$. The detail is similar to the proof of Theorem 1 of Chen and Chen (1991). Thus, the proof of Lemma 6.6.2 is finished.

According to Lemma 6.6.2, we need only to consider those A satisfying $A \subset A_0$ and $A \neq A_0$. Under the conditions of Lemma 6.6.2, we can show that there exist two constants $C_0(A)$ and $C_1(A)$ depending on A such that $h_d^* = C_0(A)T^{-1/(4+d)} \in H_T$ and for every given $A \subset \mathcal{A}$

$$\begin{aligned} CV(A) &= \inf_{h \in H_T} CV(h, A) = CV(h_d^*, A) \\ &= \widehat{\sigma}_T^2 + C_1(A)T^{-4/(4+d)} + o_P(T^{-4/(4+d)}). \end{aligned} \quad (6.6.27)$$

which implies Lemma 6.6.3. The proof of (6.6.27) is similar to that of Lemma 8 of Härdle and Vieu (1992).

Proof of Theorem 6.3.1

According to Lemma 6.6.2 again, we only need to consider those A satisfying $A \subset A_0$ and $A \neq A_0$. Thus, by (6.6.24) we have as $T \to \infty$

$$\Pr\{CV(A) > CV(A_0)\} = \Pr\Big\{C_1(A)T^{\frac{4(d-d_0)}{(4+d)(4+d_0)}} -C_1(A_0) + o_P(T^{\frac{4(d-d_0)}{(4+d)(4+d_0)}}) > 0\Big\} \to 1. \tag{6.6.28}$$

Equation (6.6.28) implies $\lim_{T\to\infty}\Pr(\widehat{A}_0 = A_0) = 1$ and therefore we complete the proof of Theorem 6.3.1.

Proof of Theorem 6.3.2

Analogous to (6.6.26), we have

$$\begin{aligned}
\widehat{\sigma}(\widehat{h}, \widehat{A}_0)^2 &= \frac{1}{T}\sum_{t=1}^{T}\{Y_t - U_{t\widehat{A}_0}^T\widehat{\beta}(\widehat{h}, \widehat{A}_0) - \widehat{g}(V_{t\widehat{A}_0}, \widehat{h})\}^2 \\
&= \frac{1}{T}\sum_{t=1}^{T} e_t^2 + \frac{1}{T}\sum_{t=1}^{T}\{U_{tA_0}^T\beta_{A_0} - U_{t\widehat{A}_0}^T\widehat{\beta}(\widehat{h}, \widehat{A}_0)\}^2 \\
&\quad + \frac{1}{T}\sum_{t=1}^{T}\{g_{A_0}(V_{tA_0}) - \widehat{g}(V_{t\widehat{A}_0}, \widehat{h})\}^2 + \frac{2}{T}\sum_{t=1}^{T}\{U_{tA_0}^T\beta_{A_0} - U_{t\widehat{A}_0}^T\widehat{\beta}(\widehat{h}, \widehat{A}_0)\}e_t \\
&\quad + \frac{2}{T}\sum_{t=1}^{T}\{U_{tA_0}^T\beta_{A_0} - U_{t\widehat{A}_0}^T\widehat{\beta}(\widehat{h}, \widehat{A}_0)\}\{g_{A_0}(V_{tA_0}) - \widehat{g}(V_{t\widehat{A}_0}, \widehat{h})\} \\
&\quad + \frac{2}{T}\sum_{t=1}^{T}\{g_{A_0}(V_{tA_0}) - \widehat{g}(V_{t\widehat{A}_0}, \widehat{h})\}e_t \\
&\equiv \frac{1}{T}\sum_{t=1}^{T} e_t^2 + \Delta(\widehat{h}, \widehat{A}_0).
\end{aligned}$$

It follows that as $T \to \infty$

$$\frac{1}{T}\sum_{t=1}^{T} e_t^2 \longrightarrow_P \sigma_0^2. \tag{6.6.29}$$

By applying Lemmas 6.6.4 and 6.6.5, it can be shown that

$$\Delta(\widehat{h}, \widehat{A}_0) \longrightarrow_P 0. \tag{6.6.30}$$

Therefore, the proof of Theorem 6.3.2 follows from (6.6.29) and (6.6.30).

Proofs of Theorems 6.4.1-6.4.5

Technical Lemmas

Lemma 6.6.6 *Assume that Assumption 6.6.8 holds. Then as $T \to \infty$*

$$\sup_x \sup_{h\in H_T} |\widehat{f}_h(x) - f(x)| = o_P(1),$$

where the $\sup_x$ *is taken over all values of* x *such that* $f(x) > 0$ *and*

$$\widehat{f}_h(\cdot) = \frac{1}{T-2}\sum_{t=3}^{T} K_h(\cdot - y_{t-2}).$$

Proof. *Lemma 6.6.6 is a weaker version of Lemma 1 of Härdle and Vieu (1992).*

Lemma 6.6.7 *(i) Assume that Assumption 6.6.8 holds. Then there exists a sequence of constants* $\{C_{ij} : 1 \le i \le 2, 1 \le j \le 2\}$ *such that*

$$\begin{aligned} L_i(h) &= \frac{1}{N}\sum_{n=1}^{N}\{\widetilde{g}_{i,h}(y_n) - g_i(y_n)\}^2 w(y_n) \\ &= C_{i1}\frac{1}{Nh} + C_{i2}h^4 + o_P(L_i(h)), \end{aligned} \tag{6.6.31}$$

$$M_i(h) = \frac{1}{N}\sum_{n=1}^{N}\{\widetilde{g}_{i,h}(y_n) - g_i(y_n)\}Z_{n+2}w(y_n) = o_P(L_i(h)) \tag{6.6.32}$$

uniformly over $h \in H_T$*, where* $Z_{n+2} = y_{n+1} - E(y_{n+1}|y_n)$.

(ii) Assume that Assumption 6.6.8 holds. Then for $i = 1, 2$

$$\sup_{h\in H_T}\frac{1}{N}\sum_{n=1}^{N}\{\widehat{g}_{i,h}(y_n) - g_i(y_n)\}^2 = o_P(N^{-1/2}), \tag{6.6.33}$$

$$\sup_{h\in H_T}\frac{1}{N}\sum_{n=1}^{N}\{\widehat{g}_{i,h}(y_n) - g_i(y_n)\}Z_{n+2} = o_P(N^{-1/2}). \tag{6.6.34}$$

Proof. (i) In order to prove Lemma 6.6.7, we need to establish the following fact

$$\begin{aligned} \widehat{g}_{i,h}(y_n) - g_i(y_n) &= \{\widehat{g}_{i,h}(y_n) - g_i(y_n)\}\frac{\widehat{f}_h(y_n)}{f(y_n)} \\ &\quad + \frac{\{\widehat{g}_{i,h}(y_n) - g_i(y_n)\}\{f(y_n) - \widehat{f}_h(y_n)\}}{f(y_n)}. \end{aligned} \tag{6.6.35}$$

Note that by Lemma 6.6.6, the second term is negligible compared to the first. Similar to the proof of Lemma 8 of Härdle and Vieu (1992), replacing Lemma 1 of Härdle and Vieu (1992) by Lemma 6.6.6, and using (6.6.35), we can obtain the proof of (6.6.31). Using the similar reason as in the proof of Lemma 2 of Härdle and Vieu (1992), we have for $i = 1, 2$

$$\bar{L}_i(h) = \frac{1}{N}\sum_{n=1}^{N}\{\widehat{g}_{i,h}(y_n) - \widehat{g}_{i,n}(y_n)\}^2 w(y_n) = o_P(L_i(h)). \tag{6.6.36}$$

It follows from (6.4.7) that for $i = 1, 2$

$$\begin{aligned} \frac{1}{N}\sum_{n=1}^{N}\{\widehat{g}_{i,h}(y_n) - g_i(y_n)\}Z_{n+2}w(y_n) &= \frac{1}{N}\sum_{n=1}^{N}\{\widehat{g}_{i,n}(y_n) - g_i(y_n)\}Z_{n+2}w(y_n) \\ &\quad + \frac{1}{N}\sum_{n=1}^{N}\{\widehat{g}_{i,h}(y_n) - \widehat{g}_{i,n}(y_n)\}Z_{n+2}w(y_n), \end{aligned}$$

where

$$\widehat{g}_{i,h}(y_n) - \widehat{g}_{i,n}(y_n) = \frac{K(0)(Nh)^{-1}\{y_{n+3-i} - \widehat{g}_{i,h}(y_n)\}}{\widehat{f}_h(y_n) - (Nh)^{-1}K(0)}. \qquad (6.6.37)$$

Observe that

$$y_{n+2} - \widehat{g}_{1,h}(y_n) = e_{n+2} + \beta u_{n+2} + g_1(y_n) - \widehat{g}_{1,h}(y_n) - \beta\{g_2(y_n) - \widehat{g}_{2,h}(y_n)\}$$

and

$$y_{n+1} - \widehat{g}_{2,h}(y_n) = u_{n+2} = Z_{n+2} + g_2(y_n) - \widehat{g}_{2,h}(y_n). \qquad (6.6.38)$$

In order to prove (6.6.32), in view of (6.6.36)–(6.6.38), it suffices to show that for $i = 1, 2$

$$\frac{1}{N}\sum_{n=1}^{N}\{\widehat{g}_{i,n}(y_n) - g_i(y_n)\}Z_{n+2}w(y_n) = o_P(L_i(h)) \qquad (6.6.39)$$

and

$$\frac{1}{N}\sum_{n=1}^{N}\{\widehat{g}_{i,h}(y_n) - \widehat{g}_{i,n}(y_n)\}Z_{n+2}w(y_n) = o_P(L_i(h)). \qquad (6.6.40)$$

The proof of (6.6.39) follows from that of (5.3) of Härdle and Vieu (1992). The proof of (6.6.40) follows from Cauchy-Schwarz inequality, (6.6.36)–(6.6.38), Lemma 6.6.6, (6.6.31) and

$$\frac{1}{N}\sum_{n=1}^{N} Z_{n+2}^2 w(y_n) = O_P(1), \qquad (6.6.41)$$

which follows from the fact that w is bounded and

$$\frac{1}{N}\sum_{n=1}^{N} Z_{n+2}^2 \to_P \sigma_2^2 \qquad (6.6.42)$$

using the standard ergodic theorem, where

$$\begin{aligned} \sigma_2^2 &= E\{y_{n+1} - E(y_{n+1}|y_n)\}^2 \\ &= Ey_{n+1}^2 - E\{E^2(y_{n+1}|y_n)\} \le 2Ey_{n+1}^2 < \infty \end{aligned}$$

using Assumption 6.6.8(vi).

(ii) Lemma 6.6.7(ii) is a weaker version of Lemma 6.6.7(i). Similar to the proof of Lemma 8 of Härdle and Vieu (1992) and the proof of (5.3) of Härdle and Vieu (1992), and using the fact that Assumption 6.6.8 still holds when the compact support of the weight function w is the same as that of the density f, we can prove both (6.6.33) and (6.6.34).

Lemma 6.6.8 *Let $\{Z_{nk}, k \geq 0\}$ be a sequence of random variables and $\{\Omega_{n,k-1}\}$ be an increasing sequence of σ-fields such that $\{Z_{nk}\}$ is measurable with respect to Ω_{nk}, and $E(Z_{nk}|\Omega_{n,k-1}) = 0$ for $1 \leq k \leq n$. If as $n \to \infty$,*

(i) $\sum_{k=1}^{n} E(Z_{nk}^2|\Omega_{n,k-1}) \to a_2^2 (> 0)$ in probability;

(ii) for every $b_2 > 0$, the sum $\sum_{k=1}^{n} E\{Z_{nk}^2 I(|Z_{nk}| > b_2)|\Omega_{n,k-1}\}$ converges in probability to zero; then as $n \to \infty$

$$\sum_{k=1}^{n} Z_{nk} \longrightarrow^{\mathcal{L}} N(0, a_2^2).$$

Proof. See Theorem 1 of Chapter VIII in Pollard (1984).

Lemma 6.6.9 *Assume that Assumption 6.6.8 holds. Then*

$$\lim_{T\to\infty} \frac{1}{T} \sup_{h\in H_T} \sum_{t=3}^{T} u_t^2 = \sigma_2^2 \tag{6.6.43}$$

in probability.

Proof. Observe that

$$\begin{aligned}\frac{1}{T}\sum_{t=3}^{T} u_t^2 &= \frac{1}{T}\sum_{t=3}^{T} Z_t^2 + \frac{1}{T}\sum_{t=3}^{T}(\bar{g}_{2t,h}^2 + 2\bar{g}_{2t,h}Z_t) \\ &\equiv \frac{1}{T}\sum_{t=3}^{T} Z_t^2 + R_T(h),\end{aligned}$$

where $\bar{g}_{2t,h} = g_2(y_{t-2}) - \widehat{g}_{2,h}(y_{t-2})$ and

$$\begin{aligned}R_T(h) &= \frac{1}{T}\sum_{t=3}^{T}\{g_2(y_{t-2}) - \widehat{g}_{2,h}(y_{t-2})\}^2 \\ &\quad + \frac{2}{T}\sum_{t=3}^{T}\{g_2(y_{t-2}) - \widehat{g}_{2,h}(y_{t-2})\}\{y_{t-1} - E(y_{t-1}|y_{t-2})\} \\ &\equiv R_{1T}(h) + R_{2T}(h).\end{aligned}$$

In view of (6.6.42), in order to prove (6.6.43), it suffices to show that for $i = 1, 2$

$$\sup_{h\in H_T} R_{iT}(h) = o_P(1),$$

which follows from Lemma 6.6.7(ii).

Proofs of Theorems 6.4.1 and 6.4.4

By (6.4.2), in order to prove Theorem 6.4.1, it suffices to show that as $T \to \infty$

$$T^{-1/2}\sum_{t=3}^{T} Z_t e_t \longrightarrow^{\mathcal{L}} N(0, \sigma^2\sigma_2^{-2}), \tag{6.6.44}$$

$$\sum_{t=3}^{T}\{\widehat{g}_{2,h}(y_{t-2}) - g_2(y_{t-2})\}e_t = o_P(T^{1/2}), \tag{6.6.45}$$

$$\sum_{t=3}^{T}\{\widehat{g}_h(y_{t-2}) - g(y_{t-2})\}Z_t = o_P(T^{1/2}), \tag{6.6.46}$$

and

$$\sum_{t=3}^{T}\{\widehat{g}_{2,h}(y_{t-2}) - g_2(y_{t-2})\}\{\widehat{g}_h(y_{t-2}) - g(y_{t-2})\} = o_P(T^{1/2}) \tag{6.6.47}$$

uniformly over $h \in H_T$, where

$$g(y_{t-2}) - \widehat{g}_h(y_{t-2}) = g_1(y_{t-2}) - \widehat{g}_{1,h}(y_{t-2}) - \beta\{g_2(y_{t-2}) - \widehat{g}_{2,h}(y_{t-2})\}. \tag{6.6.48}$$

Write Ω_{t+1} for the σ-field generated by $\{y_1, y_2; e_3, \cdots, e_t\}$. The variable $\{Z_t e_t\}$ is a martingale difference for Ω_t.

By (6.6.42) we have as $T \to \infty$

$$\frac{1}{T}\sum_{t=3}^{T} E\{(Z_t e_t)^2|\Omega_t\} \to_P \sigma^2\sigma_2^2. \tag{6.6.49}$$

Observe that for every given $b_2 > 0$

$$\begin{aligned}
&\frac{1}{T}\sum_{t=3}^{T} E\{(Z_t e_t)^2 I(|Z_t e_t| > b_2 T^{1/2})|\Omega_t\} \\
&\le \frac{1}{T}\sum_{t=3}^{T} E\{(Z_t e_t)^2 I(e_t^2 > b_2 T^{1/2})|\Omega_t\} + \frac{1}{T}\sum_{t=3}^{T} E\{(Z_t e_t)^2 I(Z_t^2 > b_2 T^{1/2})|\Omega_t\} \\
&= \frac{1}{T}\sum_{t=3}^{T} Z_t^2 E e_t^2 I(e_1^2 > b_2 T^{1/2}) + \frac{1}{T}\sum_{t=3}^{T} Z_t^2 \sigma^2 I(Z_t^2 > b_2 T^{1/2}).
\end{aligned} \tag{6.6.50}$$

Because of (6.6.42), the first sum converges to zero in probability. The second sum converges in L^1 to zero because of Assumption 6.6.8. Thus, by Lemma 6.6.8 we obtain the proof of (6.6.45).

The proofs of (6.6.45) and (6.6.46) follow from (6.6.34). The proof of (6.6.47) follows from (6.6.33) and Cauchy-Schwarz inequality. The proof of Theorem 6.4.4 follows immediately from that of Theorem 6.4.1 and the fact that $\widehat{h}_C \in H_T$ defined in (6.4.8).

Proofs of Theorems 6.4.2 and 6.4.5

In order to complete the proof of Theorem 6.4.2, we first give the following decomposition of $\widehat{\sigma}(h)^2$

$$\begin{aligned}
\widehat{\sigma}(h)^2 &= \frac{1}{T}\sum_{t=3}^{T}\{y_t - \widehat{\beta}(h)y_{t-1} - \widehat{g}_h^*(y_{t-2})\}^2 \\
&= \frac{1}{T}\sum_{t=3}^{T} e_t^2 + \frac{1}{T}\sum_{t=3}^{T} u_t^2\{\beta - \widehat{\beta}(h)\}^2 + \frac{1}{T}\sum_{t=3}^{T}\bar{g}_{t,h}^2 \\
&\quad + \frac{2}{T}\sum_{t=3}^{T} e_t u_t\{\beta - \widehat{\beta}(h)\} + \frac{2}{T}\sum_{t=3}^{T} e_t\bar{g}_{t,h} + \frac{2}{T}\sum_{t=3}^{T}\bar{g}_{t,h}u_t\{\beta - \widehat{\beta}(h)\} \\
&\equiv \sum_{j=1}^{6} J_{jh},
\end{aligned} \tag{6.6.51}$$

where $\bar{g}_{t,h} = g(y_{t-2}) - \widehat{g}_h(y_{t-2})$.

By (6.6.42) and Theorem 6.4.1 we have

$$\sup_{h\in H_T} |J_{2h}| \le o_P(T^{-1/2}). \tag{6.6.52}$$

Also, by (6.6.48) and Lemma 6.6.7(ii) we get

$$\sup_{h\in H_T} |J_{3h}| \le o_P(T^{-1/2}). \tag{6.6.53}$$

Similar to the proof of (6.6.34), we obtain for $i = 1, 2$

$$\frac{1}{T}\sum_{t=3}^{T} e_t\{\widehat{g}_{i,h}(y_{t-2}) - g_i(y_{t-2})\} \le o_P(T^{-1/2}) \tag{6.6.54}$$

uniformly over $h \in H_T$.

By Theorem 6.4.1, Lemma 6.6.7(ii), (6.6.48) and (6.6.52) we have

$$\begin{aligned}
J_{4h} &= \frac{2}{T}\{\beta - \widehat{\beta}(h)\}\Big[\sum_{t=3}^{T} e_t Z_t + \sum_{t=3}^{T} e_t\{g_2(y_{t-2}) - \widehat{g}_{2,h}(y_{t-2})\}\Big] \\
&= o_P(T^{-1/2}),
\end{aligned} \tag{6.6.55}$$

$$J_{5h} = \frac{2}{T}\sum_{t=3}^{T} e_t\{g(y_{t-2}) - \widehat{g}_h(y_{t-2})\} = o_P(T^{-1/2}), \tag{6.6.56}$$

and

$$\begin{aligned}
J_{6h} &= \frac{2}{T}\{\beta - \widehat{\beta}(h)\}\sum_{t=3}^{T} u_t\{g(y_{t-2}) - \widehat{g}_h(y_{t-2})\} \\
&= \frac{2}{T}\{\beta - \widehat{\beta}(h)\}\sum_{t=3}^{T} Z_t\{g(y_{t-2}) - \widehat{g}_h(y_{t-2})\} \\
&\quad + \frac{2}{T}\{\beta - \widehat{\beta}(h)\}\sum_{t=3}^{T} Z_t\{g(y_{t-2}) - \widehat{g}_h(y_{t-2})\}\{g_2(y_{t-2}) - \widehat{g}_{2,h}(y_{t-2})\} \\
&= o_P(T^{-1/2})
\end{aligned} \tag{6.6.57}$$

using the Cauchy-Schwarz inequality.

Therefore, the proof of Theorem 6.4.2 follows from (6.6.51)–(6.6.57) and

$$\sqrt{T}(J_{1h} - \sigma^2) \longrightarrow^{\mathcal{L}} N(0, Var(e_1^2)) \tag{6.6.58}$$

as $T \to \infty$.

The proof of Theorem 6.4.4 follows from that of Theorem 6.4.2 and the fact that $h \in H_T$ defined in (6.4.8).

Proof of Theorem 6.4.3

Before proving Theorem 6.4.3, we need to make the following notation

$$m(z_n) = \beta y_{n+1} + g(y_n), \; \widehat{m}_h(z_n) = \widehat{\beta}(h) y_{n+1} + \widehat{g}_h^*(y_n),$$

and

$$\widehat{m}_{h,n}(z_n) = \widetilde{\beta}(h) y_{n+1} + \widehat{g}_{h,n}^*(y_n), \tag{6.6.59}$$

where $z_n = (y_n, y_{n+1})^T$ and $\widehat{g}_{h,n}^*(\cdot) = \widehat{g}_{1,n}(\cdot) - \widetilde{\beta}(h)\widehat{g}_{2,n}(\cdot)$.

Observe that

$$D(h) = \frac{1}{N}\sum_{n=1}^{N}\{\widehat{m}_h(z_n) - m(z_n)\}^2 w(y_n) \tag{6.6.60}$$

and

$$CV(h) = \frac{1}{N}\sum_{n=1}^{N}\{y_{n+2} - \widehat{m}_{h,n}(z_n)\}^2 w(y_n). \tag{6.6.61}$$

In order to prove Theorem 6.4.5, noting (6.6.60) and (6.6.61), it suffices to show that

$$\sup_{h \in H_{N+2}} \frac{|\bar{D}(h) - D(h)|}{D(h)} = o_P(1) \tag{6.6.62}$$

and

$$\sup_{h, h_1 \in H_{N+2}} \frac{|\bar{D}(h) - \bar{D}(h_1) - [CV(h) - CV(h_1)]|}{D(h)} = o_P(1), \tag{6.6.63}$$

where

$$\bar{D}(h) = \frac{1}{N}\sum_{n=1}^{N}\{\widehat{m}_{h,n}(z_n) - m(z_n)\}^2 w(y_n). \tag{6.6.64}$$

We begin proving (6.6.63) and the proof of (6.6.62) is postponed to the end of this section.

Observe that the following decomposition

$$\bar{D}(h) + \frac{1}{N}\sum_{n=1}^{N} e_{n+2}^2 w(y_n) = CV(h) + 2C(h), \tag{6.6.65}$$

where

$$C(h) = \frac{1}{N}\sum_{n=1}^{N}\{\widehat{m}_{h,n}(z_n) - m(z_n)\}e_{n+2}w(y_n).$$

In view of (6.6.65), in order to prove (6.6.63), it suffices to show that

$$\sup_{h\in H_{N+2}} \frac{|C(h)|}{D(h)} = o_P(1). \tag{6.6.66}$$

Observe that

$$\begin{aligned}
C(h) = & \frac{1}{N}\sum_{n=1}^{N}\{\widehat{m}_{h,n}(z_n) - m(z_n)\}e_{n+2}w(y_n) \\
= & \frac{1}{N}\sum_{n=1}^{N} Z_{n+2}e_{n+2}w(y_n)\{\widetilde{\beta}(h) - \beta\} + \frac{1}{N}\sum_{n=1}^{N}\{\widehat{g}_{1,n}(y_n) - g_1(y_n)\}e_{n+2}w(y_n) \\
& -\{\widetilde{\beta}(h) - \beta\}\frac{1}{N}\sum_{n=1}^{N}\{\widehat{g}_{2,n}(y_n) - g_2(y_n)\}e_{n+2}w(y_n) \\
& -\beta\frac{1}{N}\sum_{n=1}^{N}\{\widehat{g}_{2,n}(y_n) - g_2(y_n)\}e_{n+2}w(y_n) \\
\equiv & \sum_{i=1}^{4} C_i(h).
\end{aligned} \tag{6.6.67}$$

Similar to the proof of Theorem 6.4.1, we can show that

$$\widetilde{\beta}(h) - \beta = O_P(N^{-1/2}) \tag{6.6.68}$$

uniformly over $h \in H_{N+2}$, where $\widetilde{\beta}(h)$ is as defined in (6.4.5).

Thus, noting that $\{Z_{n+2}e_{n+2}w(y_n)\}$ is a martingale difference, using the similar reason as the proof of (6.6.42), and applying Lemma 6.6.7(i), we have

$$\sup_{h\in H_{N+2}} \frac{|C_1(h)|}{D(h)} = o_P(1). \tag{6.6.69}$$

Therefore, noting (6.6.67)–(6.6.69), in order to prove (6.6.66), it suffices to show that for $i = 1, 2$

$$\sup_{h\in H_{N+2}} \frac{|\sum_{n=1}^{N}\{\widehat{g}_{i,n}(y_n) - g_i(y_n)\}e_{n+2}w(y_n)|}{ND(h)} = o_P(1), \tag{6.6.70}$$

which follows from Lemma 6.6.7(i) and

$$D(h) = C_1 \frac{1}{Nh} + C_2 h^4 + o_P\{D(h)\}, \tag{6.6.71}$$

where C_i $(i = 1, 2)$ are some positive constants.

Observe that

$$\begin{aligned}
D(h) &= \frac{1}{N}\sum_{n=1}^{N}\{\widehat{m}_h(z_n) - m(z_n)\}^2 w(y_n) \\
&= \frac{1}{N}\sum_{n=1}^{N}[\{\widehat{\beta}(h)y_{n+1} + \widehat{g}_{1,h}(y_n) - \widehat{\beta}(h)\widehat{g}_{2,h}(y_n)\} - \{\beta y_{n+1} + g(y_n)\}]^2 w(y_n) \\
&= \frac{1}{N}\sum_{n=1}^{N} Z_{n+2}^2 w(y_n)\{\widehat{\beta}(h) - \beta\}^2 + \frac{1}{N}\sum_{n=1}^{N}\{\widehat{g}_{1,h}(y_n) - g_1(y_n)\}^2 w(y_n) \\
&\quad + \widehat{\beta}(h)^2 \frac{1}{N}\sum_{n=1}^{N}\{\widehat{g}_{2,h}(y_n) - g_2(y_n)\}^2 w(y_n) \\
&\quad + \{\widehat{\beta}(h) - \beta\}\frac{2}{N}\sum_{n=1}^{N}\{\widehat{g}_{1,h}(y_n) - g_1(y_n)\}Z_{n+2}w(y_n) \\
&\quad - \{\widehat{\beta}(h) - \beta\}\widehat{\beta}(h)\frac{2}{N}\sum_{n=1}^{N}\{\widehat{g}_{2,h}(y_n) - g_2(y_n)\}Z_{n+2}w(y_n) \\
&\quad - \widehat{\beta}(h)\frac{2}{N}\sum_{n=1}^{N}\{\widehat{g}_{2,h}(y_n) - g_2(y_n)\}\{\widehat{g}_{1,h}(y_n) - g_1(y_n)\}w(y_n) \\
&\equiv \sum_{i=1}^{6} D_i(h).
\end{aligned} \tag{6.6.72}$$

Using (6.6.41) and Theorem 6.4.1, we have

$$\sup_{h \in H_{N+2}} |D_1(h)| = O_P(N^{-1}). \tag{6.6.73}$$

By Theorem 6.4.1 and Lemma 6.6.7(i), we obtain for $i = 2, 3$

$$D_i(h) = C_{i1}\frac{1}{Nh} + C_{i2}h^4 + o_P\{D_i(h)\}, \tag{6.6.74}$$

$$D_4(h) = o_P(D_2(h)) \text{ and } D_5(h) = o_P(D_3(h)), \tag{6.6.75}$$

where C_{i1} and C_{i2} are real positive constants.

Therefore, applying Cauchy-Schwarz inequality to $D_6(h)$ and using (6.6.31) again, we complete the proof of (6.6.71).

We now prove (6.6.62). By the definition of $\bar{D}(h)$, we have

$$\bar{D}(h) = \frac{1}{N}\sum_{n=1}^{N}\{\widehat{m}_{h,n}(z_n) - m(z_n)\}^2 w(y_n)$$

$$
\begin{aligned}
&= \frac{1}{N}\sum_{n=1}^{N}[\{\widetilde{\beta}(h)y_{n+1}+\widehat{g}_{1,n}(y_n)-\widetilde{\beta}(h)\widehat{g}_{2,n}(y_n)\}-\{\beta y_{n+1}+g(y_n)\}]^2 w(y_n)\\
&= \frac{1}{N}\sum_{n=1}^{N} Z_{n+2}^2 w(y_n)(\widetilde{\beta}(h)-\beta)^2+\frac{1}{N}\sum_{n=1}^{N}\{\widehat{g}_{1,n}(y_n)-g_1(y_n)\}^2 w(y_n)\\
&\quad +\widetilde{\beta}(h)^2\frac{1}{N}\sum_{n=1}^{N}\{\widehat{g}_{2,n}(y_n)-g_2(y_n)\}^2 w(y_n)\\
&\quad +\{\widetilde{\beta}(h)-\beta\}\frac{2}{N}\sum_{n=1}^{N}\{\widehat{g}_{1,n}(y_n)-g_1(y_n)\}Z_{n+2}w(y_n)\\
&\quad -\{\widetilde{\beta}(h)-\beta\}\widetilde{\beta}(h)\frac{2}{N}\sum_{n=1}^{N}\{\widehat{g}_{2,n}(y_n)-g_2(y_n)\}Z_{n+2}w(y_n)\\
&\quad -\widetilde{\beta}(h)\frac{2}{N}\sum_{n=1}^{N}\{\widehat{g}_{2,n}(y_n)-g_2(y_n)\}\{\widehat{g}_{1,n}(y_n)-g_1(y_n)\}w(y_n)\\
&\equiv \sum_{i=1}^{6}\bar{D}_i(h). \qquad (6.6.76)
\end{aligned}
$$

Firstly, by (6.6.72) and (6.6.76) we obtain

$$
\bar{D}_1(h)-D_1(h)=\frac{1}{N}\sum_{n=1}^{N} Z_{n+2}^2 w(y_n)\{\widetilde{\beta}(h)-\widehat{\beta}(h)\}[\{\widetilde{\beta}(h)-\beta\} +\{\widehat{\beta}(h)-\beta\}]. \qquad (6.6.77)
$$

Similar to the proof of Theorem 6.4.1, we can prove

$$
\widetilde{\beta}(h)-\widehat{\beta}(h)=o_P(N^{-1/2}), \qquad (6.6.78)
$$

Thus, (6.6.41) and Theorem 6.4.1 imply

$$
\sup_{h\in H_{N+2}}\frac{|\bar{D}_1(h)-D_1(h)|}{D(h)}=o_P(1). \qquad (6.6.79)
$$

Secondly, by the similar reason as the proof of Lemma 2 of Härdle and Vieu (1992) we have for $i=1,2$

$$
\sup_{h\in H_{N+2}}\frac{|\Delta\bar{D}_i(h)|}{D(h)}=o_P(1), \qquad (6.6.80)
$$

where

$$
\Delta\bar{D}_i(h)=\frac{1}{N}\sum_{n=1}^{N}[\{\widehat{g}_{i,n}(y_n)-g_i(y_n)\}^2-\{\widehat{g}_{i,h}(y_n)-g_i(y_n)\}^2]w(y_n).
$$

Thus

$$
\sup_{h\in H_{N+2}}\frac{|\bar{D}_2(h)-D_2(h)|}{D(h)}=o_P(1). \qquad (6.6.81)
$$

On the other hand, observe that

$$\bar{D}_3(h) - D_3(h) = \widetilde{\beta}(h)^2 \Delta \bar{D}_2(h) + \{\widetilde{\beta}(h) - \widehat{\beta}(h)\}\{\widetilde{\beta}(h) + \widehat{\beta}(h)\} D_3(h). \quad (6.6.82)$$

Hence, by Theorem 6.4.1, (6.6.68), (6.6.74), (6.6.78), and (6.6.80) we get

$$\sup_{h \in H_{N+2}} \frac{|\bar{D}_3(h) - D_3(h)|}{D(h)} = o_P(1). \quad (6.6.83)$$

Thirdly, by (6.6.32) and (6.6.36)–(6.6.41), we obtain for $i = 4, 5$

$$\sup_{h \in H_{N+2}} \frac{|\bar{D}_i(h) - D_i(h)|}{D(h)} = o_P(1), \quad (6.6.84)$$

where

$$\begin{aligned} \bar{D}_4(h) - D_4(h) &= \{\widehat{\beta}(h) - \beta\} \frac{2}{N} \sum_{n=1}^{N} \{\widehat{g}_{1,n}(y_n) - \widehat{g}_{1,h}(y_n)\} Z_{n+2} w(y_n) \\ &\quad + \{\widetilde{\beta}(h) - \widehat{\beta}(h)\} \frac{2}{N} \sum_{n=1}^{N} \{\widehat{g}_{1,n}(y_n) - g_1(y_n)\} Z_{n+2} w(y_n) \end{aligned}$$

and

$$\begin{aligned} \bar{D}_5(h) - D_5(h) &= -\{\widehat{\beta}(h) - \beta\} \widehat{\beta}(h) \frac{2}{N} \sum_{n=1}^{N} \{\widehat{g}_{2,n}(y_n) - \widehat{g}_{2,h}(y_n)\} Z_{n+2} w(y_n) \\ &\quad - \{\widetilde{\beta}(h) - \widehat{\beta}(h)\}\{\widetilde{\beta}(h) + \widehat{\beta}(h) - \beta\} \\ &\quad \times \frac{2}{N} \sum_{n=1}^{N} \{\widehat{g}_{2,n}(y_n) - g_2(y_n)\} Z_{n+2} w(y_n). \end{aligned} \quad (6.6.85)$$

Finally, note that the following decomposition

$$\begin{aligned} \bar{D}_6(h) \quad - \quad & D_6(h) = -\frac{2\widetilde{\beta}(h)}{N} \Big[\sum_{n=1}^{N} \{\widehat{g}_{1,n}(y_n) - \widehat{g}_{1,h}(y_n)\}\{\widehat{g}_{2,n}(y_n) - \widehat{g}_{2,h}(y_n)\} \\ & + \sum_{n=1}^{N} \{\widehat{g}_{1,h}(y_n) - g_1(y_n)\}\{\widehat{g}_{2,n}(y_n) - \widehat{g}_{2,h}(y_n)\} \\ & + \sum_{n=1}^{N} \{\widehat{g}_{1,n}(y_n) - \widehat{g}_{1,h}(y_n)\}\{\widehat{g}_{2,h}(y_n) - g_2(y_n)\} \Big] w(y_n) \\ & + \frac{2\{\widetilde{\beta}(h) - \widehat{\beta}(h)\}}{N} \sum_{n=1}^{N} \{\widehat{g}_{1,h}(y_n) - g_1(y_n)\}\{\widehat{g}_{2,h}(y_n) - g_2(y_n)\} w(y_n) \\ \equiv \quad & E_1(h) + E_2(h) + E_3(h) + E_4(h). \end{aligned} \quad (6.6.86)$$

Thus, applying Cauchy-Schwarz inequality, using Lemma 6.6.7(i) and equations (6.6.36)–(6.6.41), we have for $i = 1, 2, 3, 4$

$$\sup_{h \in H_{N+2}} \frac{|E_i(h)|}{D(h)} = o_P(1). \quad (6.6.87)$$

Therefore, the proof of (6.6.62) follows from (6.6.76)–(6.6.87).

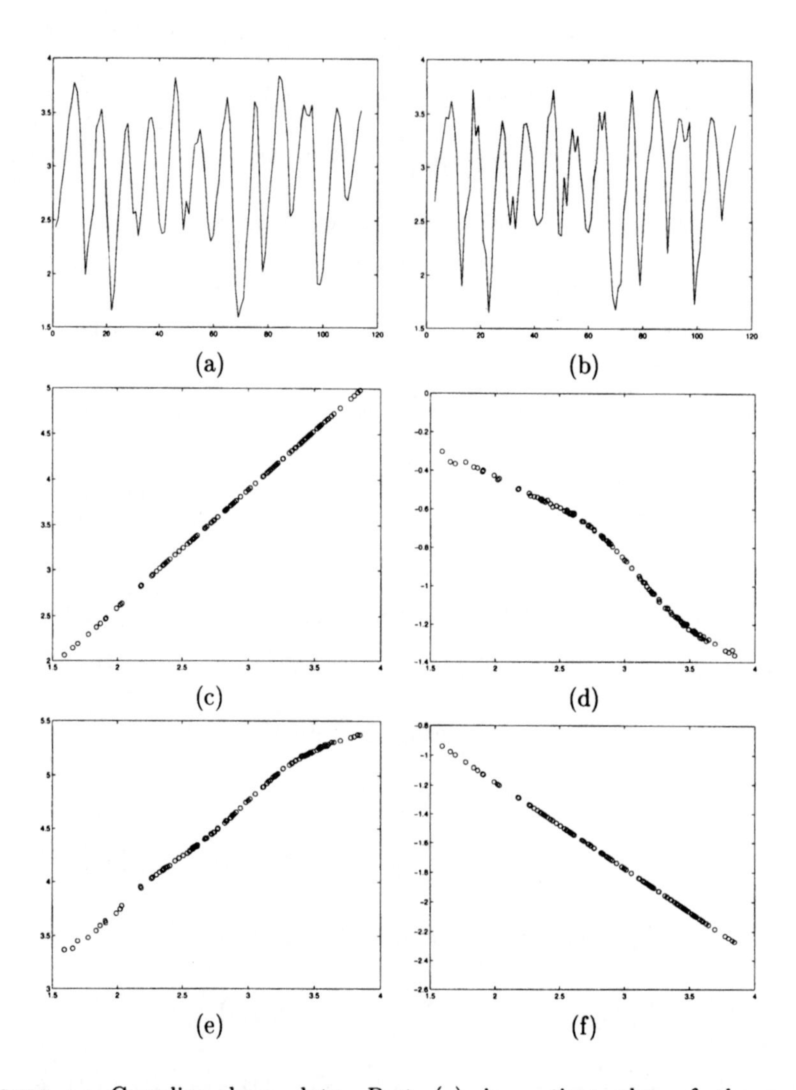

FIGURE 6.1. Canadian lynx data. Part (a) is a time plot of the common-log-transformed lynx data. Part (b) is a plot of fitted values (solid) and the observations (dashed). Part (c) is a plot of the LS estimate against the first lag for model (6.4.22). Part (d) is a plot of the nonparametric estimate against the second lag for model (6.4.22). Part (e) is a plot of the nonparametric estimate against the first lag for model (6.4.23) and part (f) is a plot of the LS estimate against the second lag for model (6.4.23).

APPENDIX: BASIC LEMMAS

In this appendix, we state several famous results including Abel's inequality and Bernstein's inequality and then prove some lemmas which are often used in the previous chapters.

Abel's Inequality. *Let $\{\xi_n, n \geq 1\}$ be a sequence of real numbers such that $\sum \xi_n$ converges, and let $\{\eta_n\}$ be a monotone decreasing sequence of positive constants. Then*

$$\eta_1\Big(\min_{1\leq k\leq n}\sum_{i=1}^{k}\xi_i\Big) \leq \sum_{i=1}^{n}\xi_i\eta_i \leq \eta_1\Big(\max_{1\leq k\leq n}\sum_{i=1}^{k}\xi_i\Big).$$

Bernstein's Inequality. *Let $V_1, \ldots, V_n$ be independent random variables with zero means and bounded ranges: $|V_i| \leq M$. Then for each $\eta > 0$,*

$$P(|\sum_{i=1}^{n}V_i| > \eta) \leq 2\exp\Big[-\eta^2/\Big\{2\Big(\sum_{i=1}^{n}\mathrm{Var}V_i + M\eta\Big)\Big\}\Big].$$

Lemma A.1 *Suppose that Assumptions 1.3.1 and 1.3.3 (iii) hold. Then*

$$\max_{1\leq i\leq n}|G_j(T_i) - \sum_{k=1}^{n}\omega_{nk}(T_i)G_j(T_k)| = O(c_n) \quad \text{for } j = 0, \ldots, p,$$

where $G_0(\cdot) = g(\cdot)$ and $G_l(\cdot) = h_l(\cdot)$ for $l = 1, \ldots, p$.

Proof. We only present the proof for $g(\cdot)$. The proofs of the other cases are similar. Observe that

$$\begin{aligned}
\sum_{i=1}^{n}\omega_{ni}(t)g(T_i) - g(t) &= \sum_{i=1}^{n}\omega_{ni}(t)\{g(T_i) - g(t)\} + \Big\{\sum_{i=1}^{n}\omega_{ni}(t) - 1\Big\}g(t) \\
&= \sum_{i=1}^{n}\omega_{ni}(t)\{g(T_i) - g(t)\}I(|T_i - t| > c_n) \\
&\quad + \sum_{i=1}^{n}\omega_{ni}(t)\{g(T_i) - g(t)\}I(|T_i - t| \leq c_n) \\
&\quad + \Big\{\sum_{i=1}^{n}\omega_{ni}(t) - 1\Big\}g(t).
\end{aligned}$$

By Assumption 1.3.3(b) and Lipschitz continuity of $g(\cdot)$,

$$\sum_{i=1}^{n}\omega_{ni}(t)\{g(T_i) - g(t)\}I(|T_i - t| > c_n) = O(c_n) \tag{A.1}$$

and

$$\sum_{i=1}^{n} \omega_{ni}(t)\{g(T_i) - g(t)\}I(|T_i - t| \le c_n) = O(c_n). \tag{A.2}$$

(A.1)-(A.2) and Assumption 1.3.3 complete our proof.

Lemma A.2 *If Assumptions 1.3.1-1.3.3 hold. Then*

$$\lim_{n\to\infty} n^{-1}\widetilde{\mathbf{X}}^T\widetilde{\mathbf{X}} = \Sigma.$$

Proof. Denote $\overline{h}_{ns}(T_i) = h_s(T_i) - \sum_{k=1}^{n} \omega_{nk}(T_i)X_{ks}$. It follows from $X_{js} = h_s(T_j) + u_{js}$ that the (s, m) element of $\widetilde{\mathbf{X}}^T\widetilde{\mathbf{X}}$ $(s, m = 1, \ldots, p)$ can be decomposed as:

$$\begin{aligned} & \sum_{j=1}^{n} u_{js}u_{jm} + \sum_{j=1}^{n} \overline{h}_{ns}(T_j)u_{jm} + \sum_{j=1}^{n} \overline{h}_{nm}(T_j)u_{js} + \sum_{j=1}^{n} \overline{h}_{ns}(T_j)\overline{h}_{nm}(T_j) \\ \stackrel{\text{def}}{=} & \sum_{j=1}^{n} u_{js}u_{jm} + \sum_{q=1}^{3} R_{nsm}^{(q)}. \end{aligned}$$

The strong laws of large number imply that $\lim_{n\to\infty} 1/n \sum_{i=1}^{n} u_i u_i^T = \Sigma$ and Lemma A.1 means $R_{nsm}^{(3)} = o(n)$. Using `Cauchy-Schwarz inequality` and the above arguments, we can show that $R_{nsm}^{(1)} = o(n)$ and $R_{nsm}^{(2)} = o(n)$. This completes the proof of the lemma.

As mentioned above, Assumption 1.3.1 (i) holds when (X_i, T_i) are i.i.d. random design points. Thus, Lemma A.2 holds with probability one when (X_i, T_i) are i.i.d. random design points.

Next we shall prove a general result on strong uniform convergence of weighted sums, which is often applied in the monograph. See Liang (1999) for its proof.

Lemma A.3 *(Liang, 1999) Let $V_1, \ldots, V_n$ be independent random variables with $EV_i = 0$ and finite variances, and $\sup_{1\le j\le n} E|V_j|^r \le C < \infty$ $(r \ge 2)$. Assume that $\{a_{ki}, k, i = 1\ldots, n\}$ is a sequence of real numbers such that $\sup_{1\le i,k\le n} |a_{ki}| = O(n^{-p_1})$ for some $0 < p_1 < 1$ and $\sum_{j=1}^{n} a_{ji} = O(n^{p_2})$ for $p_2 \ge \max(0, 2/r - p_1)$. Then*

$$\max_{1\le i\le n}\Big|\sum_{k=1}^{n} a_{ki}V_k\Big| = O(n^{-s}\log n) \quad \textit{for} \quad s = (p_1 - p_2)/2, \quad a.s.$$

Lemma A.3 considers the case where $\{a_{ki}, k, i = 1\ldots, n\}$ is a sequence of real numbers. As a matter of fact, the conclusion of Lemma A.3 remains unchanged when $\{a_{ki}, k, i = 1\ldots, n\}$ is a sequence of random variables satisfying

$\sup_{1\le i,k\le n}|a_{ki}| = O(n^{-p_1})$ a.s. and $\sum_{j=1}^n a_{ji} = O(n^{p_2})$ a.s. for some $0 < p_1 < 1$ and $p_2 \ge \max(0, 2/r - p_1)$. Thus, we have the following useful results: Let $r = 3$, $V_k = e_k$ or u_{kl}, $a_{ji} = \omega_{nj}(T_i)$, $p_1 = 2/3$ and $p_2 = 0$. We obtain the following formulas, which plays critical roles throughout the monograph.

$$\max_{i\le n}\Big|\sum_{k=1}^n \omega_{nk}(T_i)e_k\Big| = O(n^{-1/3}\log n), \quad a.s. \tag{A.3}$$

and

$$\max_{i\le n}\Big|\sum_{k=1}^n \omega_{nk}(T_i)u_{kl}\Big| = O(n^{-1/3}\log n) \quad \text{for } l = 1, \ldots, p. \tag{A.4}$$

Lemma A.4 *Let $V_1, \ldots, V_n$ be independent random variables with $EV_i = 0$ and $\sup_{1\le i\le n} EV_i^4 < \infty$. Assume that Assumption 1.3.3 with $b_n = Cn^{-3/4}(\log n)^{-1}$ holds. Then for n large enough, we have $I_n = o(n^{1/2})$ a.s., where*

$$I_n = \sum_{i=1}^n \sum_{j\ne i} \omega_{nj}(T_i)(V_j' - EV_j')(V_i' - EV_i'),$$

in which $V_i' = V_i I(|V_i| \le i^{1/4})$.

Proof. See Gao (1995a) for its proof.

REFERENCES

Akahira, M. & Takeuchi, K.(1981). *Asymptotic Efficiency of Statistical Estimators. Concepts and Higher Order Asymptotic Efficiency.* Lecture Notes in Statistics **7**, Springer-Verlag, New York.

An, H. Z. & Huang, F. C. (1996). The geometrical ergodicity of nonlinear autoregressive models. *Statistica Sinica*, **6**, 943–956.

Andrews, D. W. K. (1991). Asymptotic normality of series estimates for nonparametric and semiparametric regression models. *Econometrica*, **59**, 307–345.

Anglin, P.M. & Gencay, R. (1996). Semiparametric estimation of a hedonic price function. *Journal of Applied Econometrics*, **11**, 633-648.

Bahadur, R.R. (1967). Rates of convergence of estimates and test statistics. *Ann. Math. Stat.*, **38**, 303-324.

Begun, J. M., Hall, W. J., Huang, W. M. & Wellner, J. A. (1983). Information and asymptotic efficiency in parametric-nonparametric models. *Annals of Statistics*, **11**, 432-452. ALS, **14**, 1295–1298.

Bhattacharya, P.K. & Zhao, P. L.(1997). Semiparametric inference in a partial linear model. *Annals of Statistics*, **25**, 244-262.

Bickel, P.J. (1978). Using residuals robustly I: Tests for heteroscedasticity, nonlinearity. *Annals of Statistics*, **6**, 266-291.

Bickel, P. J. (1982). On adaptive estimation. *Annals of Statistics*, **10**, 647-671.

Bickel, P. J., Klaasen, C. A. J., Ritov, Ya'acov & Wellner, J. A. (1993). *Efficient and Adaptive Estimation for Semiparametric Models.* The Johns Hopkins University Press.

Blanchflower, D.G. & Oswald, A.J. (1994). *The Wage Curve*, MIT Press Cambridge, MA.

Boente, G. & Fraiman, R. (1988). Consistency of a nonparametric estimator of a density function for dependent variables. *Journal of Multivariate Analysis*, **25**, 90–99.

Bowman, A. & Azzalini,A. (1997). *Applied Smoothing Techniques: the Kernel Approach and S-plus Illustrations.* Oxford University Press, Oxford.

Box, G. E. P. & Hill, W. J. (1974). Correcting inhomogeneity of variance with power transformation weighting. *Technometrics*, **16**, 385-389.

Buckley, M. J. & Eagleson, G. K. (1988). An approximation to the distribution of quadratic forms in normal random variables. *Australian Journal of Statistics*, **30A**, 150–159.

Buckley, J. & James, I. (1979). Linear regression with censored data. *Biometrika*, **66**, 429-436.

Carroll, R. J. (1982). Adapting for heteroscedasticity in linear models. *Annals of Statistics*, **10**, 1224-1233.

Carroll, R. J. & Härdle, W. (1989). Second order effects in semiparametric weighted least squares regression. *Statistics*, **2**, 179-186.

Carroll, R. J. & Ruppert, D. (1982). Robust estimation in heteroscedasticity linear models. *Annals of Statistics*, **10**, 429-441.

Carroll, R. J., Ruppert, D. & Stefanski, L. A. (1995). *Nonlinear Measurement Error Models*. Vol. 63 of *Monographs on Statistics and Applied Probability*, Chapman and Hall, New York.

Carroll, R. J., Fan, J., Gijbels, I. & Wand, M. P. (1997). Generalized partially single-index models. *Journal of the American Statistical Association*, **92**, 477-489.

Chai, G. X. & Li, Z. Y. (1993). Asymptotic theory for estimation of error distributions in linear model. *Science in China, Ser. A*, **14**, 408-419.

Chambers, J. M. & Hastie, T. J. (ed.) (1992). *Statistical Models in S.* Chapman and Hall, New York.

Chen, H.(1988). Convergence rates for parametric components in a partly linear model. *Annals of Statistics*, **16**, 136-146.

Chen, H. & Chen, K. W. (1991). Selection of the splined variables and convergence rates in a partial linear model. *The Canadian Journal of Statistics*, **19**, 323-339.

Chen, R. & Tsay, R. (1993). Nonlinear additive ARX models. *Journal of the American Statistical Association*, **88**, 955–967.

Chen, R., Liu, J. & Tsay, R. (1995). Additivity tests for nonlinear autoregression. *Biometrika*, **82**, 369–383.

Cheng, B. & Tong, H. (1992). On consistent nonparametric order determination and chaos. *Journal of the Royal Statistical Society, Series B*, **54**, 427–449.

Cheng, B. & Tong, H. (1993). Nonparametric function estimation in noisy chaos. *Developments in time series analysis.* (In honour of Maurice B. Priestley, ed. T. Subba Rao), 183–206. Chapman and Hall, London.

Cheng, P. (1980). Bahadur asymptotic efficiency of MLE. *Acta Mathematica Sinica*, **23**, 883-900.

Cheng, P., Chen, X., Chen, G.J & Wu, C.Y. (1985). *Parameter Estimation.* Science & Technology Press, Shanghai.

Chow, Y. S. & Teicher, H. (1988). *Probability Theory.* 2nd Edition, Springer-Verlag, New York.

Cuzick, J. (1992a). Semiparametric additive regression. *Journal of the Royal Statistical Society, Series B*, **54**, 831-843.

Cuzick, J. (1992b). Efficient estimates in semiparametric additive regression models with unknown error distribution. *Annals of Statistics*, **20**, 1129-1136.

Daniel, C. & Wood, F. S. (1971). *Fitting Equations to Data.* John Wiley, New York.

Daniel, C. & Wood, F. S. (1980). *Fitting Equations to Data: Computer Analysis of Multifactor Data for Scientists and Engineers.* 2nd Editor. John Wiley, New York.

Davison, A. C. & Hinkley, D. V. (1997). *Bootstrap Methods and their Application.* Cambridge University Press, Cambridge.

Devroye, L. P. & Wagner, T.J. (1980). The strong uniform consistency of kernel estimates. *Journal of Multivariate Analysis*, **5**, 59-77.

DeVore, R. A. & Lorentz, G. G. (1993). *Constructive Approximation.* New York: Springer.

Dinse, G.E. & Lagakos, S.W. (1983). Regression analysis of tumour prevalence. *Applied Statistics*, **33**, 236-248.

Doukhan, P. (1995). *Mixing: Properties and Examples.* Lecture Notes in Statistics, **86**, Springer, New York.

Efron, B. & Tibshirani, R. J. (1993). *An Introduction to the Bootstrap.* Vol. 57 of *Monographs on Statistics and Applied Probability*, Chapman and Hall, New York.

Engle, R. F., Granger, C. W. J., Rice, J. & Weiss, A. (1986). Semiparametric estimates of the relation between weather and electricity sales. *Journal of the American Statistical Association*, **81**, 310-320.

Eubank, R.L. (1988). *Spline Smoothing and Nonparametric Regression.* New York, Marcel Dekker.

Eubank, R.L., Kambour,E.L., Kim,J.T., Klipple, K., Reese C.S. and Schimek, M. (1998). Estimation in partially linear models. *Computational Statistics & Data Analysis*, **29**, 27-34.

Eubank, R. L. & Spiegelman, C. H. (1990). Testing the goodness-of-fit of a linear model via nonparametric regression techniques. *Journal of the American Statistical Association*, **85**, 387–392.

Eubank, R. L. & Whitney, P. (1989). Convergence rates for estimation in certain partially linear models. *Journal of Statistical Planning & Inference*, **23**, 33-43.

Fan, J. & Truong, Y. K. (1993). Nonparametric regression with errors in variables. *Annals of Statistics*, **21**, 1900–1925.

Fan, J. & Gijbels, I. (1996). *Local Polynomial Modelling and Its Applications.* Vol. 66 of *Monographs on Statistics and Applied Probability*, Chapman and Hall, New York.

Fan, Y. & Li, Q. (1996). Consistent model specification tests: omitted variables and semiparametric functional forms. *Econometrica*, **64**, 865–890.

Fu, J. C. (1973). On a theorem of Bahadur on the rate of convergence of point estimations. *Annals of Statistics*, **1**, 745-749.

Fuller, W. A. (1987). *Measurement Error Models.* John Wiley, New York.

Fuller, W. A. & Rao, J. N. K. (1978). Estimation for a linear regression model with unknown diagonal covariance matrix. *Annals of Statistics*, **6**, 1149-1158.

Gallant, A. R. (1981). On the bias in flexible functional forms and an essentially unbiased form: the Fourier flexible form. *J. Econometrics*, **15**, 211–245.

Gao, J. T. (1992). Large Sample Theory in Semiparametric Regression Models. Ph.D. Thesis, Graduate School, University of Science & Technology of China, Hefei, P.R. China.

Gao, J. T. (1995a). The laws of the iterated logarithm of some estimates in partly linear models. *Statistics & Probability Letters*, **25**, 153-162.

Gao, J. T. (1995b). Asymptotic theory for partly linear models. *Communications in Statistics, Theory & Methods*, **24**, 1985-2010.

Gao, J. T. (1998). Semiparametric regression smoothing of nonlinear time series. *Scandinavian Journal of Statistics*, **25**, 521-539.

Gao, J. T. & Anh, V. V. (1999). Semiparametric regression under long-range dependent errors. *Journal of Statistical Planning & Inference*, **80**, 37-57.

Gao, J. T., Chen, X. R. & Zhao, L. C.(1994). Asymptotic normality of a class estimates in partly linear models. *Acta Mathematica Sinica*, **37**, 256-268.

Gao, J. T., Hong, S.Y. & Liang, H. (1995). Convergence rates of a class of estimates in partly linear models. *Acta Mathematica Sinica*, **38**, 658-669.

Gao, J. T., Hong, S. Y. & Liang, H. (1996). The Berry-Esseen bounds of parametric component estimation in partly linear models. *Chinese Annals of Mathematics*, **17**, 477-490.

Gao, J. T. & Liang, H. (1995). Asymptotic normality of pseudo-LS estimator for partially linear autoregressive models. *Statistics & Probability Letters*, **23**, 27–34.

Gao, J. T. & Liang, H. (1997). Statistical inference in single-index and partially nonlinear regression models. *Annals of the Institute of Statistical Mathematics*, **49**, 493–517.

Gao, J. T. & Shi, P. D. (1997). M-type smoothing splines in nonparametric and semiparametric regression models. *Statistica Sinica*, **7**, 1155–1169.

Gao, J. T., Tong, H. & Wolff, R. (1998a). Adaptive series estimation in additive stochastic regression models. Technical report No. 9801, School of Mathematical Sciences, Queensland University of Technology, Brisbane, Australia.

Gao, J. T., Tong, H. & Wolff, R. (1998b). Adaptive testing for additivity in additive stochastic regression models. Technical report No. 9802, School of Mathematical Sciences, Queensland University of Technology, Brisbane, Australia.

Gao, J. T. & Zhao, L. C. (1993). Adaptive estimation in partly linear models. *Sciences in China Ser. A*, **14**, 14-27.

Glass, L. & Mackey, M. (1988). *From Clocks to Chaos: the Rhythms of Life*. Princeton University Press, Princeton.

Gleser, L. J. (1992). The importance of assessing measurement reliability in multivariate regression. *Journal of the American Statistical Association*, **87**, 696-707.

Golubev, G. & Härdle, W. (1997). On adaptive in partial linear models. *Discussion paper no. 371, Weierstrsse-Institut für Angewandte Analysis und Stochastik zu Berlin.*

Green, P., Jennison, C. & Seheult, A. (1985). Analysis of field experiments by least squares smoothing. *Journal of the Royal Statistical Society, Series B*, **47**, 299-315.

Green, P. & Silverman, B. W. (1994). *Nonparametric Regression and Generalized Linear Models: a Roughness Penalty Approach*. Vol. 58 of *Monographs on Statistics and Applied Probability*, Chapman and Hall, New York.

Green, P. & Yandell, B. (1985). Semi-parametric generalized linear models. *Generalized Linear Models* (R. Gilchrist, B. J. Francis and J. Whittaker, eds), Lecture Notes in Statistics, **30**, 44-55. Springer, Berlin.

GSOEP (1991). *Das Sozio-ökonomische Panel (SOEP) im Jahre 1990/91*, Projektgruppe "Das Sozio-ökonomische Panel", Deutsches Institut für Wirtschaftsforschung. Vierteljahreshefte zur Wirtschaftsforschung, pp. 146–155.

Gu, M. G. & Lai, T. L. (1990). Functional laws of the iterated logarithm for the product-limit estimator of a distribution function under random censorship or truncated. *Annals of Probability*, **18**, 160-189.

Györfi, L., Härdle, W., Sarda, P. & Vieu, P. (1989). *Nonparametric curve estimation for time series.* Lecture Notes in Statistics, **60**, Springer, New York.

Hall, P. & Carroll, R. J. (1989). Variance function estimation in regression: the effect of estimating the mean. *Journal of the Royal Statistical Society, Series B*, **51**, 3-14.

Hall, P. & Heyde, C. C. (1980). *Martingale Limit Theory and Its Applications.* Academic Press, New York.

Hamilton, S. A. & Truong, Y. K. (1997). Local linear estimation in partly linear models. *Journal of Multivariate Analysis*, **60**, 1-19.

Härdle, W. (1990). *Applied Nonparametric Regression.* Cambridge University Press, New York.

Härdle, W. (1991). *Smoothing Techniques: with Implementation in S.* Springer, Berlin.

Härdle, W., Klinke, S. & Müller, M. (1999). *XploRe Learning Guide.* Springer-Verlag.

Härdle, W., Lütkepohl, H. & Chen, R. (1997). A review of nonparametric time series analysis. *International Statistical Review*, **21**, 49–72.

Härdle, W. & Mammen, E. (1993). Testing parametric versus nonparametric regression. *Annals of Statistics*, **21**, 1926–1947.

Härdle, W., Mammen, E. & Müller, M. (1998). Testing parametric versus semiparametric modeling in generalized linear models. *Journal of the American Statistical Association*, **93**, 1461-1474.

Härdle, W. & Vieu, P. (1992). Kernel regression smoothing of time series. *Journal of Time Series Analysis*, **13**, 209–232.

Heckman, N.E. (1986). Spline smoothing in partly linear models. *Journal of the Royal Statistical Society, Series B*, **48**, 244-248.

Hjellvik, V. & Tjøstheim, D. (1995). Nonparametric tests of linearity for time series. *Biometrika*, **82**, 351–368.

Hong, S. Y. (1991). Estimation theory of a class of semiparametric regression models. *Sciences in China Ser. A*, **12**, 1258-1272.

Hong, S. Y. & Cheng, P.(1992a). Convergence rates of parametric in semiparametric regression models. Technical report, Institute of Systems Science, Chinese Academy of Sciences.

Hong, S. Y. & Cheng, P.(1992b). The Berry-Esseen bounds of some estimates in semiparametric regression models. Technical report, Institute of Systems Science, Chinese Academy of Sciences.

Hong, S. Y. & Cheng, P. (1993). Bootstrap approximation of estimation for parameter in a semiparametric regression model. *Sciences in China Ser. A*, **14**, 239-251.

Hong, Y. & White, H. (1995). Consistent specification testing via nonparametric series regression. *Econometrica*, **63**, 1133–1159.

Jayasuriva, B. R. (1996). Testing for polynomial regression using nonparametric regression techniques. *Journal of the American Statistical Association*, **91**, 1626–1631.

Jobson, J. D. & Fuller, W. A. (1980). Least squares estimation when covariance matrix and parameter vector are functionally related. *Journal of the American Statistical Association*, **75**, 176-181.

Kashin, B. S. & Saakyan, A. A. (1989). *Orthogonal Series.* Translations of Mathematical Monographs, **75**.

Koul, H., Susarla, V. & Ryzin, J. (1981). Regression analysis with randomly right-censored data. *Annals of Statistics*, **9**, 1276-1288.

Kreiss, J. P., Neumann, M. H. & Yao, Q. W. (1997). Bootstrap tests for simple structures in nonparametric time series regression. Private communication.

Lai, T. L. & Ying, Z. L. (1991). Rank regression methods for left-truncated and right-censored data. *Annals of Statistics*, **19**, 531-556.

Lai, T. L. & Ying, Z. L. (1992). Asymptotically efficient estimation in censored and truncated regression models. *Statistica Sinica*, **2**, 17-46.

Liang, H. (1992). Asymptotic Efficiency in Semiparametric Models and Related Topics. Thesis, Institute of Systems Science, Chinese Academy of Sciences, Beijing, P.R. China.

Liang, H. (1994a). On the smallest possible asymptotically efficient variance in semiparametric models. *System Sciences & Matematical Sciences*, **7**, 29-33.

Liang, H. (1994b). The Berry-Esseen bounds of error variance estimation in a semiparametric regression model. *Communications in Statistics, Theory & Methods*, **23**, 3439-3452.

Liang, H. (1995a). On Bahadur asymptotic efficiency of maximum likelihood estimator for a generalized semiparametric model. *Statistica Sinica*, **5**, 363-371.

Liang, H. (1995b). Second order asymptotic efficiency of PMLE in generalized linear models. *Statistics & Probability Letters*, **24**, 273-279.

Liang, H. (1995c). A note on asymptotic normality for nonparametric multiple regression: the fixed design case. *Soo-Chow Journal of Mathematics*, 395-399.

Liang, H. (1996). Asymptotically efficient estimators in a partly linear autoregressive model. *System Sciences & Matematical Sciences*, **9**, 164-170.

Liang, H. (1999). An application of Bernstein's inequality. *Econometric Theory*, **15**, 152.

Liang, H. & Cheng, P. (1993). Second order asymptotic efficiency in a partial linear model. *Statistics & Probability Letters*, **18**, 73-84.

Liang, H. & Cheng, P. (1994). On Bahadur asymptotic efficiency in a semiparametric regression model. *System Sciences & Matematical Sciences*, **7**, 229-240.

Liang, H. & Härdle, W. (1997). Asymptotic properties of parametric estimation in partially linear heteroscedastic models. *Technical report no 33*, Sonderforschungsbereich 373, Humboldt-Universität zu Berlin.

Liang, H. & Härdle, W. (1999). Large sample theory of the estimation of the error distribution for semiparametric models. *Communications in Statistics, Theory & Methods*, in press.

Liang, H., Härdle, W. & Carroll, R.J. (1999). Estimation in a semiparametric partially linear errors-in-variables model. *Annals of Statistics*, in press.

Liang, H., Härdle, W. & Sommerfeld, V. (1999). Bootstrap approximation of the estimates for parameters in a semiparametric regression model. *Journal of Statistical Planning & Inference*, in press.

Liang, H., Härdle, W. & Werwatz, A. (1999). Asymptotic properties of nonparametric regression estimation in partly linear models. *Econometric Theory*, **15**, 258.

Liang, H. & Huang, S.M. (1996). Some applications of semiparametric partially linear models in economics. *Science Decision*, **10**, 6-16.

Liang, H. & Zhou, Y. (1998). Asymptotic normality in a semiparametric partial linear model with right-censored data. *Communications in Statistics, Theory & Methods*, **27**, 2895-2907.

Linton, O.B. (1995). Second order approximation in the partially linear regression model. *Econometrica*, **63**, 1079-1112.

Lu, K. L. (1983). On Bahadur Asymptotically Efficiency. Dissertation of Master Degree. Institute of Systems Science, Chinese Academy of Sciences.

Mak, T. K. (1992). Estimation of parameters in heteroscedastic linear models. *Journal of the Royal Statistical Society, Series B*, **54**, 648-655.

Mammen, E. & van de Geer, S. (1997). Penalized estimation in partial linear models. *Annals of Statistics*, **25**, 1014-1035.

Masry, E. & Tjøstheim, D. (1995). Nonparametric estimation and identification of nonlinear ARCH time series. *Econometric Theory*, **11**, 258–289.

Masry, E. & Tjøstheim, D. (1997). Additive nonlinear ARX time series and projection estimates. *Econometric Theory*, **13**, 214–252.

Müller, H. G. & Stadtmüller, U. (1987). Estimation of heteroscedasticity in regression analysis. *Annals of Statistics*, **15**, 610-625.

Müller, M. & Rönz, B. (2000). *Credit scoring using semiparametric methods*, in Springer LNS (eds. J. Franks, W. Härdle and G. Stahl).

Nychka, D., Elliner, S., Gallant, A. & McCaffrey, D. (1992). Finding chaos in noisy systems. *Journal of the Royal Statistical Society, Series B*, **54**, 399-426.

Pollard, D. (1984). *Convergence of Stochastic Processes.* Springer, New York.

Rendtel, U. & Schwarze, J. (1995). Zum Zusammenhang zwischen Lohnhoehe und Arbeitslosigkeit: Neue Befunde auf Basis semiparametrischer Schaetzungen und eines verallgemeinerten Varianz-Komponenten Modells. German Institute for Economic Research (DIW) Discussion Paper 118, Berlin, 1995.

Rice, J.(1986). Convergence rates for partially splined models. *Statistics & Probability Letters*, **4**, 203-208.

Robinson, P.M.(1983). Nonparametric estimation for time series models. *Journal of Time Series Analysis*, **4**, 185-208.

Robinson, P.M.(1988). Root-n-consistent semiparametric regression. *Econometrica*, **56**, 931-954.

Schick, A. (1986). On asymptotically efficient estimation in semiparametric model. *Annals of Statistics*, **14**, 1139-1151.

Schick, A. (1993). On efficient estimation in regression models. *Annals of Statistics*, **21**, 1486-1521. (Correction and Addendum, **23**, 1862-1863).

Schick, A. (1996a). Weighted least squares estimates in partly linear regression models. *Statistics & Probability Letters*, **27**, 281-287.

Schick, A. (1996b). Root-n consistent estimation in partly linear regression models. *Statistics & Probability Letters*, **28**, 353-358.

Schimek, M. (1997). Non- and semiparametric alternatives to generalized linear models. *Computational Statistics.* **12**, 173-191.

Schimek, M. (1999). Estimation and inference in partially linear models with smoothing splines. *Journal of Statistical Planning & Inference*, to appear.

Schmalensee, R. & Stoker, T.M. (1999). Household gasoline demand in the United States. *Econometrica*, **67**, 645-662.

Severini, T. A. & Staniswalis, J. G. (1994). Quasilikelihood estimation in semiparametric models. *Journal of the American Statistical Association*, **89**, 501-511.

Speckman, P. (1988). Kernel smoothing in partial linear models. *Journal of the Royal Statistical Society, Series B*, **50**, 413-436.

Stefanski, L. A. & Carroll, R. (1990). Deconvoluting kernel density estimators. *Statistics*, **21**, 169-184.

Stone, C. J. (1975). Adaptive maximum likelihood estimation of a location parameter. *Annals of Statistics*, **3**, 267-284.

Stone, C. J. (1982). Optimal global rates of convergence for nonparametric estimators. *Annals of Statistics*, **10**, 1040-1053.

Stone, C. J. (1985). Additive regression and other nonparametric models. *Annals of Statistics*, **13**, 689-705.

Stout, W. (1974). *Almost Sure Convergence.* Academic Press, New York.

Teräsvirta, T., Tjøstheim, D. & Granger, C. W. J. (1994). Aspects of modelling nonlinear time series, in R. F. Engle & D. L. McFadden (eds.). *Handbook of Econometrics*, **4**, 2919–2957.

Tjøstheim, D. (1994). Nonlinear time series: a selective review. *Scandinavian Journal of Statistics*, **21**, 97–130.

Tjøstheim, D. & Auestad, B. (1994a). Nonparametric identification of nonlinear time series: projections. *Journal of the American Statistical Association*, **89**, 1398-1409.

Tjøstheim, D. & Auestad, B. (1994b). Nonparametric identification of nonlinear time series: selecting significant lags. *Journal of the American Statistical Association*, **89**, 1410–1419.

Tong, H. (1977). Some comments on the Canadian lynx data (with discussion). *Journal of the Royal Statistical Society, Series A*, **140**, 432–436.

Tong, H. (1990). *Nonlinear Time Series.* Oxford University Press.

Tong, H. (1995). A personal overview of nonlinear time series analysis from a chaos perspective (with discussions). *Scandinavian Journal of Statistics*, **22**, 399–445.

Tripathi, G. (1997). Semiparametric efficiency bounds under shape restrictions. Unpublished manuscript, Department of Economics, University of Wisconsin-Madison.

Vieu, P. (1994). Choice of regression in nonparametric estimation. *Computational Statistics & Data Analysis*, **17**, 575-594.

Wand, M. & Jones, M.C. (1994). *Kernel Smoothing*. Vol. 60 of *Monographs on Statistics and Applied Probability*, Chapman and Hall, London.

Whaba, G. (1990). *Spline Models for Observational Data, CBMS-NSF Regional Conference Series in Applied Mathematics*. **59**, Philadelphia, PA: SIAM, XII.

Willis, R. J. (1986). *Wage Determinants: A Survey and Reinterpretation of Human Capital Earnings Functions in: Ashenfelter, O. and Layard, R. The Handbook of Labor Economics*, Vol.1 North Holland-Elsevier Science Publishers Amsterdam, 1986, pp 525-602.

Wong, C. M. & Kohn, R. (1996). A Bayesian approach to estimating and forecasting additive nonparametric autoregressive models. *Journal of Time Series Analysis*, **17**, 203–220.

Wu, C. F. J. (1986). Jackknife, Bootstrap and other resampling methods in regression analysis (with discussion). *Annals of Statistics*, **14**, 1261-1295.

Yao, Q. W. & Tong, H. (1994). On subset selection in nonparametric stochastic regression. *Statistica Sinica*, **4**, 51–70.

Zhao, L. C. (1984). Convergence rates of the distributions of the error variance estimates in linear models. *Acta Mathematica Sinica*, **3**, 381-392.

Zhao, L. C. & Bai, Z. D. (1985). Asymptotic expansions of the distribution of sum of the independent random variables. *Sciences in China Ser.* A, **8**, 677-697.

Zhou, M. (1992). Asymptotic normality of the synthetic data regression estimator for censored survival data. *Annals of Statistics*, **20**, 1002-1021.

AUTHOR INDEX

SUBJECT INDEX

SYMBOLS AND NOTATION

The following notation is used throughout the monograph.

a.s.	almost surely
i.i.d.	independent and identically distributed
$\mathcal{F}$	the identity matrix of order p
CLT	central limit theorem
LIL	law of the iterated logarithm
MLE	maximum likelihood estimate
$\text{Var}(\xi)$	the variance of ξ
$N(a, \sigma^2)$	normal distribution with mean a and variance σ^2
$U(a, b)$	uniform distribution on (a, b)
$\stackrel{\text{def}}{=}$	denote
$\longrightarrow^{\mathcal{L}}$	convergence in distribution
$\longrightarrow_P$	convergence in probability
S^T	the transpose of vector or matrix S
$\mathbf{X}$	$(X_1, \ldots, X_n)^T$
$\mathbf{Y}$	$(Y_1, \ldots, Y_n)^T$
$\mathbf{T}$	$(T_1, \ldots, T_n)^T$
$\omega_{nj}(\cdot)$ or $\omega^*_{nj}(\cdot)$	weight functions
$\widetilde{\mathbf{S}}$	$(\widetilde{S}_1, \ldots, \widetilde{S}_n)^T$ with $\widetilde{S}_i = S_i - \sum_{j=1}^n \omega_{nj}(T_i)S_j$, where S_i represents a random variable or a function.
$\widetilde{\mathbf{G}}$	$(\widetilde{g}_1, \ldots, \widetilde{g}_n)^T$ with $\widetilde{g}_i = g(T_i) - \sum_{j=1}^n \omega_{nj}(T_i)g(T_j)$.
$\xi_n = O_p(\eta_n)$	for any $\zeta > 0$, there exist M and n_0 such that $P\{\lvert\xi_n\rvert \geq M\lvert\eta_n\rvert\} < \zeta$ for any $n \geq n_0$
$\xi_n = o_p(\eta_n)$	$P\{\lvert\xi_n\rvert \geq \zeta\lvert\eta_n\rvert\} \to 0$ for each $\zeta > 0$
$\xi_n = o_p(1)$	ξ_n converges to zero in probability
$O_p(1)$	stochastically bounded
$S^{\otimes 2}$	SS^T
$S^{-1} = (s^{ij})_{p\times p}$	the inverse of $S = (s_{ij})_{p\times p}$
$\Phi(x)$	standard normal distribution function
$\phi(x)$	standard normal density function

For convenience and simplicity, we always let C denote some positive constant which may have different values at each appearance throughout this monograph.

Druck: Strauss Offsetdruck, Mörlenbach
Verarbeitung: Schäffer, Grünstadt

LaVergne, TN USA
02 April 2011
222632LV00003B/121/A